Lecture Notes in Mathematics

Edited by A. Dold and B. Eckmann

599

Complex Analysis

Proceedings of the Conference held at the University of Kentucky, May 18–22, 1976

Edited by
J. D. Buckholtz and T. J. Suffridge

Springer-Verlag
Berlin · Heidelberg · New York 1977

Editors
James D. Buckholtz
Teddy J. Suffridge
Department of Mathematics
University of Kentucky
Lexington, KY 40506/USA

Library of Congress Cataloging in Publication Data

Conference on Complex Analysis, University of Kentucky, 1976.
Complex analysis.

(Lecture notes in mathematics ; 599)
1. Functions of complex variables—Congresses.
2. Shah, S. M., 1905- I. Buckholtz, James D., 1935- II. Suffridge, T. J. III. Title.
IV. Series: Lecture notes in mathematics (Berlin) ; 599.
QA3.L28 no. 599 [QA331] 510'.8s [515'.9] 77-9883

AMS Subject Classifications (1970): 30A08, 30A15, 30A20, 30A36, 30A40, 30A42, 30A60, 30A64, 30A70, 32A15, 32A30

ISBN 3-540-08343-X Springer-Verlag Berlin · Heidelberg · New York
ISBN 0-387-08343-X Springer-Verlag New York · Heidelberg · Berlin

Printing and binding: Beltz Offsetdruck, Hemsbach/Bergstr.
2141/3140-543210

PREFACE

This volume consists of papers that were presented at the conference on Complex Analysis held at the University of Kentucky during May 18–22, 1976. This conference was held in honor of Professor S.M. Shah on the occasion of his retirement. He has contributed important work in various branches of the broad area of complex analysis including papers on entire and meromorphic functions, univalent functions and analytic number theory. It is therefore appropriate that the topics considered in this volume are varied although there is a certain unity in that the problems considered have a geometric motivation or interpretation.

The conference received generous support from the National Science Foundation under Grant Number MCS 76-06305-A01 and from the University of Kentucky Graduate School. We wish to express our appreciation for this support.

We thank all participants for their cooperation and for making the conference a success. We also thank our colleagues Professors Frank Keogh and John L. Lewis who did much of the work necessary for organizing the conference. Finally, we wish to thank Linda Newton and Karen Shepherd who attended to many details of the conference and Pat Nichols who did an excellent job of typing the manuscript.

J. D. Buckholtz

T. J. Suffridge

Lexington, Kentucky

October, 1976

Swarupchand M. Shah

Professor of Mathematics, University of Kentucky

Professor Shah is a quiet, hard-working mathematician of rare ability and knowledge. He is an excellent teacher with the unusual ability to cover a great amount of material in a meaningful way and still have time to answer questions that are raised in class. His kind and helpful manner has attracted many students at both the master's and Ph.D. levels. Since 1964, he has directed 18 Ph.D. theses. He is never too busy to help a student.

Professor Shah's broad knowledge and wide range of interests is evidenced by the fact that he has published papers and taught courses in Complex Analysis, Fourier Series, Approximation Theory and Theory of Numbers. He has directed theses in most, if not all, of these areas. He has published over 130 research papers in a variety of well-known mathematical journals as well as co-authoring the texts: (i) Introduction to Real Variable Theory, with S. C. Saxena, Intext, 1972 and (ii) Summability Theory and Applications, with R. E. Powell, Van Nostrand, 1972 and three other books prior to 1958.

He is a native of India and was born December 30, 1905. He attended the colleges affiliated to the University of Bombay during 1923–27 and received the degree B.A. (Honors) from that University. He continued his education at the University College, University of London during 1927–30 and received the following degrees from that institution: M.A. 1930; Ph.D. 1942; D.Litt. 1951. His major advisor was E. C. Titchmarsh.

His professional career includes positions at Muslim University, Aligarh as Senior Lecturer, Reader and finally Professor and Department Chairman over a period of 28 years. He was a Government of India Scholar at Oxford University and at Cambridge University. He came to the United States in 1958 and held visiting positions at the University of Wisconsin, Northwestern University and the University of Kansas before accepting the position Professor of Mathematics at the University of Kentucky in 1966.

Professor Shah has received a number of honors among which are a listing in World Who's Who in Science, Who's Who in India and American Men of Science. He is a fellow of the Royal Society of Edinburgh, Scotland and a fellow of the Indian National Science Academy, New Delhi. In 1969, he was given the highest honor that can be bestowed by his colleagues at the University of Kentucky in that he was named Distinguished Professor in the College of Arts and Sciences. Also, in 1969, he was honored by the University of Kentucky Research Foundation and was awarded a prize for outstanding research. He has been editor and associate editor of a number of professional journals and has reviewed extensively for Scripta Mathematica, Mathematical Reviews and Zentralblatt. He has been invited to lecture at many institutions both in the U.S. and abroad including U.C.L.A., Stanford, University of Aberdeen, Scotland, Universite de Montreal and many other institutions. In addition, he gave an address at the 1969 meeting of the American Mathematical Society in Baton Rouge, Louisiana and an address at the 1974 Spring meeting of the Missouri Section of the Mathematical Association of America.

I am certain that Professor Shah will continue to be a productive mathematician as a Professor Emeritus at the University of Kentucky. It will be a pleasure to continue to have him as a colleague at the University and to be able to seek his advice and counsel although, as one of his former students, I am still somewhat awed by his ability and knowledge.

Ted Suffridge

TABLE OF CONTENTS

CONFERENCE PARTICIPANTS

L. V. Ahlfors	Harvard University
Albert Baernstein II	Washington University, St. Louis
Roger Barnard	Texas Tech. University
James Brennan	University of Kentucky
Johnny Brown	University of Michigan
J. D. Buckholtz	University of Kentucky
Emanuel Calys	Washburn University, Topeka, Kansas
William Causey	University of Mississippi
Joseph Cima	University of North Carolina
Björn Dahlberg	University of Göteborg, Sweden
M. Damodaran	Purdue University
David Drasin	Purdue University
Peter Duren	University of Michigan
Albert Edrei	Syracuse University
Doug Elosser	University of Kentucky
Michael Freeman	University of Kentucky
Gerd Fricke	Wright State University, Dayton
W. H. J. Fuchs	Cornell University
Ron Gariepy	University of Kentucky
A. W. Goodman	University of South Florida
Fred Gross	University of Maryland, Baltimore
Lawrence Harris	University of Kentucky
Lowell Hansen	Wayne State University
W. K. Hayman	Imperial College, London
Larry Heath	University of Texas, Arlington
Maurice Heins	University of Maryland
Simon Hellerstein	University of Wisconsin
F. R. Keogh	University of Kentucky
Amy King	Eastern Kentucky University
Larry Kotman	Purdue University
Eric P. Kronstadt	University of Michigan
Jackson Lackey	University of Kentucky
George Leeman	I.B.M. Corporation (New York)
Yuk Leung	University of Michigan
John Lewis	University of Kentucky
John Mack	University of Kentucky
Ed Merkes	University of Cincinnati
Joseph Miles	University of Illinois
Sanford Miller	S.U.N.Y., Brockport
R. E. Pippert	Purdue University
George Piranian	University of Michigan
Robert Powell	Kent State University
Charles Rees	University of Louisiana, New Orleans
Ray Roan	University of Michigan

W. C. Royster	University of Kentucky
Lee Rubel	University of Illinois
Walter Rudin	University of Wisconsin
Ed Saff	University of South Florida
Mohammad Salmassi	University of Kentucky
Glenn Schober	Indiana University
I. J. Schoenberg	M.R.C. University of Wisconsin
S. M. Shah	University of Kentucky
J. K. Shaw	Virginia Tech.
Daniel Shea	University of Wisconsin
Linda Sons	Northern Illinois University
T. J. Suffridge	University of Kentucky
Selden Trimble	University of Missouri, Rolla
Anna Tsao	University of Michigan
Mark Vernon	University of Kentucky
Barnet Weinstock	University of Kentucky
Allen Weitsman	Purdue University
John Wermer	Brown University
Donald Wright	University of Cincinnati
Jang Mei Wy	University of Illinois
C. C. Yang	Naval Research Laboratory

A SOMEWHAT NEW APPROACH TO QUASICONFORMAL MAPPINGS IN R^n

Lars V. Ahlfors

1. Quasiconformal mappings are differentiable almost everywhere. We shall denote by DF the Jacobian matrix of a q.c. mapping F from one region in R^n to another, and for convenience we shall denote its determinant by $|DF|$ (we assume all mappings to be sense-preserving so that $|DF| > 0$ a.e.). It is also useful to introduce a very neutral notation for the *normalized Jacobian*, for instance,

$$XF = |DF|^{-1/n} DF \quad . \tag{1}$$

The nature of the dilatation at a point is best described by the symmetrized matrix

$$MF = (XF)^T XF \quad . \tag{2}$$

It is, in a way, the analogue of the complex dilatation $f_{\bar{z}}/f_z$ in two dimensions.

Notations. We use A^T for the transpose, $\operatorname{tr} A$ for the trace of a square matrix, and we define the norm $||A||$ by $||A||^2 = \operatorname{tr} A^T A$.

2. Because MF is positive definite, it can be written in the form

$$MF = U^T \Lambda^2 U \, , \tag{3}$$

where $U \in O(n)$ and $\Lambda = [\lambda_1, \ldots, \lambda_n]$ is a diagonal matrix with positive entries. We choose the notation so that $\lambda_1 \geqslant \cdots \geqslant \lambda_n$. Observe that $\lambda_1 \cdots \lambda_n = 1$.

It follows from (2) and (3) that

$$XF = V \Lambda U \, , \tag{4}$$

where V is another orthogonal matrix.

3. There are various ways of measuring the amount of quasiconformality, or the deviation from conformality. The most natural thought is to use the essential supremum of λ_1/λ_n, but in their basic paper [5] Gehring and Väisälä have given good reasons for introducing *two* coefficients of quasiconformality, namely

$$K_0(F) = \text{ess sup}\, \lambda_1 \quad , \qquad K_I(F) = \text{ess sup}\, \lambda_n^{-1} \quad .$$

Another possibility would be to use the supremum of $||XF||$, or rather $n^{-1/2}||XF||$, in order to characterize conformal mappings by the number 1. Note that $||XF||^2 = \lambda_1^2 + \cdots + \lambda_n^2$.

There is still another choice, suggested to me by Earle (unpublished). The space of positive definite matrices Y can be made into a Riemannian manifold by introducing the metric

$$ds(Y)^2 = \operatorname{tr}\,(Y^{-1}dY\, Y^{-1}dy) \tag{5}$$

which is obviously invariant in the sense that it does not change when Y is replaced by $Y[A] = A^T YA$.

The distance between Y_1 and Y_2 is now defined as the length in the metric (5) of the shortest arc joining Y_1 and Y_2. In particular, we set

$$d(M) = \text{dist}(I,M) = \text{dist}(I,\Lambda^2)$$

where I is the unit matrix and M is short for MF. The shortest path is seen to be $Y(t) = \Lambda^{2t}$, $t \in [0,1]$. One finds

$$d(M) = [\sum_\alpha (\log \lambda_\alpha^2)^2]^{1/2} . \tag{6}$$

If $h(\lambda)$ is any reasonable function it is customary to write $h(\Lambda^2) = [h(\lambda_1^2), \ldots, h(\lambda_n^2)]$ and $h(M) = U^T h(\Lambda^2) U$ for M as in (3). With this notation we have simply

$$d(M) = \|\log M\| . \tag{7}$$

Earle's suggestion is to use ess sup $d(MF)$, or a multiple of the same, as coefficient of quasiconformality. It is the same for a mapping F and its inverse F^{-1}, and it is subadditive under composition.

4. It is worthwhile to compare with the case $n = 2$. At any given point, the Jacobian can be written as

$$DF = r(\theta') \begin{pmatrix} 1+k & 0 \\ 0 & 1-k \end{pmatrix} r(\theta) ,$$

where $r(\theta)$, $r(\theta')$ are rotations and $0 \leqslant k < 1$. It follows that

$$MF = r(-\theta) \begin{pmatrix} K & 0 \\ 0 & K^{-1} \end{pmatrix} r(-\theta)$$

with the customary notation $K = (1+k)/(1-k)$. One obtains $d(M) = \sqrt{2} \log K$. Recall that the supremum of $\log K$ is used when defining the classical Teichmüller distance.

For arbitrary n it is natural to normalize the "logarithmic coefficient of quasiconformality" in such a way that the affine mapping which takes x_1 to Kx_1 and leaves the other coordinates fixed has coefficient $\log K$. For this mapping $d(M) = 2\left(\frac{n-1}{n}\right)^{1/2} \log K$. We are, therefore, suggesting the following terminology:

Definition. F is K-q.c. in the sense of Earle-Teichmüller if $\|\log MF\| \leqslant 2\left(\frac{n-1}{n}\right)^{1/2} \log K$ almost everywhere.

5. The first impression is undoubtedly that the proposed definition is likely to be difficult to handle technically. A crucial test would be whether, with this definition, the affine mapping is extremal for the classical rectangular box problem of Grötsch. At the moment I have not yet been able to decide, conclusively, whether this is true or not.

On the other hand, the specific motivation for this paper is rather encouraging. It is my purpose to show that it is surprisingly easy to calculate the variation of $\|\log MF\|$ that corresponds to a small deformation.

For simplicity, we treat only the case of self-mappings of the unit ball $B = \{x \in R^n \mid |x| < 1\}$.

A deformation of B is a continuous mapping $f\colon B \to R^n$. The infinitesimal version of MF is

$$(8) \qquad Sf = \tfrac{1}{2}(Df + Df^T) - \frac{1}{n}(\mathrm{tr}\, Df)I\,,$$

where the derivatives are assumed to exist as locally integrable distributional derivatives. If, in addition, $\|Sf\|$ is bounded, for instance $\|Sf\| \leqslant k$ a.e., then f is said to be a quasiconformal deformation.

The boundedness of $\|Sf\|$ has a strong influence on the regularity of f. It can be shown (see [3]) that f satisfies a condition of the form

$$(9) \qquad |f(x) - f(y)| = O(|x - y| \log \frac{1}{|x - y|})$$

for small $|x - y|$, uniformly on every compact set. Actually, we shall assume that f is a deformation of the closed unit ball on itself. This means that f has continuous boundary values which satisfy the orthogonality condition $f \perp x$ on $|x| = 1$. Under this condition f can be extended by symmetry to all of R^n, and (9) will hold uniformly on the unit ball.

6. The deformation f generates a one-parameter group of homeomorphisms $F(x,t)$ given as solutions of the differential equations

$$(10) \qquad \dot{F}(x,t) = f(F(x,t)), \qquad F(x,0) = x$$

where the dot stands for differentiation with respect to t. Local existence is trivial, and (9) is an Osgood condition which guarantees uniqueness. The boundary condition forces $F(x,t)$ to stay inside B, and this in turn implies existence for all time. Each F_t, defined by $F_t(x) = F(x,t)$, is a homeomorphism, and $F_{s+t} = F_s \circ F_t$, $F_t^{-1} = F_{-t}$. All this is very well known.

We wish to prove:

Theorem 1. *If* $\|Sf\| \leqslant k$, *then* $\|\log MF_t\| \leqslant 2k|t|$.

In other words, if f is a quasiconformal deformation, then each F_t is a quasiconformal mapping whose degree of quasiconformality stays within predictable bounds. A similar theorem was first proved by Reimann [7] who used different techniques and different norms.

7. In the proof that follows we are going to assume that f is of class C^1. The passage to the general case is quite easy and follows the pattern of the proof of Theorem 5 in [3].

We begin by determining $(MF_t)^{\cdot}$. For brevity we shall write X and M for XF_t and MF_t, $\dot{X}$ and $\dot{M}$ for their derivatives. From (10) we have at once $(DF_t)^{\cdot} = (Df \circ F_t)DF_t$ and by the familar rule for differentiating a determinant $|DF_t|^{\cdot} = |DF_t|\,\mathrm{tr}(Df \circ F_t)$. Applied to (1) and (2) this gives[1]

$$(11) \qquad \begin{aligned} \dot{X} &= (Df \circ F_t - \frac{1}{n}\mathrm{tr}\, Df \circ F_t)X \\ \dot{M} &= 2X^T(Sf \circ F_t)X\,. \end{aligned}$$

[1] Here and later we follow the practice of denoting a numerical multiple cI by c.

8. Next we have to find the derivative of $\|\log M\|^2 = \Sigma_\alpha (\log \lambda_\alpha^2)^2$. For this purpose we choose a contour C in the complex plane which encloses all the λ_α^2, but not 0, and evaluate the sum as a residue integral. The λ_α^2 are the zeros of $\det(zI - M)$ whose logarithmic derivative is $\operatorname{tr}(zI - M)^{-1}$. We obtain

(12) $$\|\log M\|^2 = \frac{1}{2\pi i}\int_C \operatorname{tr}(zI - M)^{-1}(\log z)^2\, dz$$

where $\log z$ is real for $z > 0$. This formula can be differentiated at once and yields

(13) $$(\|\log M\|^2)^{\cdot} = \frac{1}{2\pi i}\int_C \operatorname{tr}[(zI - M)^{-2}\dot{M}](\log z)^2\, dz\ .$$

The integrand in (13) has double poles at the λ_α^2. In fact, the explicit expression for the trace is

$$\Sigma_{\alpha ij}\, U_{\alpha i}(z - \lambda_\alpha^2)^{-2}\, U_{\alpha j}\, \dot{M}_{ij}\ .$$

With the notation adapted in Sec. 3, it follows that

(14) $$(\|\log M\|^2)^{\cdot} = 2\operatorname{tr}(M^{-1}\log M\, \dot{M})$$

and finally, substituting from (11),

(15) $$(\|\log M\|^2)^{\cdot} = 4\operatorname{tr}[XM^{-1}\log M\, X^T(Sf \circ F_t)]\ .$$

By the Cauchy-Schwarz inequality and the assumption $\|Sf\| \leqslant k$ the right hand side of (15) is $\leqslant 4k\|XM^{-1}\log M\, X^T\|$. However, $X^TX = M$ implies $\|XM^{-1}\log M\, X^T\| = \|\log M\|$. The resulting differential inequality leads at once to $\|\log M\| \leqslant 2k|t|$ as asserted.

9. As a second application of the same technique, we consider the problem of extremal quasiconformal mappings of the unit ball on itself with prescribed boundary correspondence. The problem can be presented as follows: Let $\Phi: B \to B$ be a given quasiconformal mapping of the unit ball. It is well known that Φ extends to a homeomorphism of the boundary. When is Φ *extremal* in the sense that ess sup $\|\log M\Phi\| \leqslant$ ess sup $\|\log M\Phi_1\|$ for any Φ_1 that agrees with Φ on the boundary?

To simplify the notation we shall write ess sup $\|\ \| = \|\ \|_{2,\infty}$. This should remind the reader that we are referring to the uniform norm of the square norm of the matrix.

We shall prove the following necessary condition:

Theorem 2. *Suppose that Φ is extremal, and let f be a quasiconformal deformation with $f(x) = 0$ for $|x| = 1$. Then*

(16) $$\|\log M\Phi\|_{2,\infty} \leqslant \|\log M\Phi - Sf\|_{2,\infty}\ .$$

Proof. Let f be as assumed and form F_t as in Sec. 6. It is clear that $F_t(x) = x$ on the boundary of B. Therefore, $\Phi_t = \Phi \circ F_t^{-1}$ has the same boundary values as Φ.

The chain rule implies $X\Phi_t \circ F_t = X\Phi(XF_t)^{-1}$ and hence

(17) $$M\Phi_t \circ F_t = (X_t^T)^{-1}\, M\Phi X_t^{-1}$$

where $X_t = XF_t$. We differentiate this identity at $t = 0$, observing that $X_0 = I$ while according to (11) $\dot{X}_0 = Df - \frac{1}{n} \operatorname{tr} Df$. It follows that

$$[(X_t^T)^{-1} M\Phi X_t^{-1}]^{\cdot}_{t=0} = -M\Phi(Df - \frac{1}{n} \operatorname{tr} Df) - (Df - \frac{1}{n} \operatorname{tr} Df)^T M\Phi .$$

We substitute this expression for $\dot{M}$ in (14) and obtain

(18) $$(\|\log M\Phi_t \circ F_t\|^2)^{\cdot}_{t=0} = -4 \operatorname{tr}[(\log M\Phi)Sf] .$$

Accordingly,

$$\|\log M\Phi_t \circ F_t\|^2 = \|\log M\Phi\|^2 - 4t \operatorname{tr}[(\log M\Phi)Sf] + o(t)$$

$$\leqslant (1-4t)\|\log M\Phi\|^2 + 4t\|\log M\Phi\| \cdot \|\log M\Phi - Sf\| + o(t)$$

for $t > 0$. This implies

(19) $$\|\log M\Phi_t\|^2_{2,\infty} \leqslant (1-4t)\|\log M\Phi\|^2_{2,\infty} + 4t\|\log M\Phi\|_{2,\infty} \cdot \|\log M\Phi - Sf\|_{2,\infty} + o(t) .$$

If (16) were false, the right hand side of (19) would be strictly less than $\|\log M\Phi\|^2_{2,\infty}$ for small t, thereby contradicting the extremality of Φ. This reasoning is borrowed from Bers [4].

10. There is another interpretation of Theorem 2 which makes it a generalization of (part of) Hamilton's necessary condition for extremal q.c. mappings in the plane (Hamilton [6], Bers [4]). Let us denote by $L^1(B)$ the space of (measurable) matrix valued functions $s(x)$ with norm

$$\|s\|_{2,1} = \int_B \|s\| \, dx < \infty .$$

The bounded linear functionals on $L^1(B)$ are of the form

$$\langle A,s \rangle = \int_B \operatorname{tr}(As) \, dx$$

where A is a symmetric matrix function and $\|A\|_{2,\infty} < \infty$. The operator norm of A is precisely $\|A\|_{2,\infty}$.

Let Q be a linear subspace of $L^1(B)$. The operator norm of A acting on Q is denoted by $\|A\|_Q$, i.e. $\|A\|_Q = \sup |\langle A,s \rangle|$ for $s \in Q$, $\|s\| \leqslant 1$. It is clear that $\|A\|_Q \leqslant \|A\|_{2,\infty}$.

In the work of Hamilton, Q is the space of integrable quadratic analytic differentials. In our adaptation we shall let Q consist of all $\sigma \in L^1(B)$ which satisfy the following additional conditions:

(i) $\operatorname{tr} \sigma = 0$

(ii) $\Sigma_i D_i \sigma_{ij} = 0, \; j = 1, \ldots, n$.

11. Let A be symmetric with $\|A\|_{2,\infty} < \infty$. With the above choice of Q we define:

Definition. A is Hamiltonian if $\|A\|_Q = \|A\|_{2,\infty}$.

Theorem 2 turns out to be equivalent to the following:

Theorem 2a. *If* Φ *is extremal, then* $\log M\Phi$ *is Hamiltonian.*

The proof requires somewhat elaborate preparations and will not be given in this paper.

References

1. L. V. Ahlfors, Conditions for quasiconformal deformations in several variables, in Contributions to Analysis, A Collection of Papers Dedicated to Lipman Bers, Acad. Press, New York and London, 1974, 19–25.

2. ____________, Invariant operators and integral representations in hyperbolic space, *Mathematica Scandinavica* **36** (1975), 27–43.

3. ____________, Quasiconformal deformations and mappings in R^n, to appear in *Journal d'Analyse.*

4. L. Bers, Extremal quasiconformal mappings, *Annals of Mathematics Studies* **66** (1971), 27–52.

5. F. W. Gehring, J. Väisälä, The coefficients of quasiconformality of domains in space, *Acta Math.* **114** (1965), 1–70.

6. R. S. Hamilton, Extremal quasiconformal mappings with prescribed boundary values, *Trans. Am. Math. Soc.* **138** (1969), 399–406.

7. H. M. Reimann, Ordinary differential equations and quasiconformal mappings, to appear in *Advances in Mathematics.*

Division of Applied Mathematics
Harvard University
Cambridge, Massachusetts 02138

AN INFINITE ORDER PERIODIC ENTIRE FUNCTION WHICH IS PRIME

I. N. Baker and Chung-Chun Yang

An entire function $F(z)$ is said to have a factorization with left factor $f(z)$ and right factor $g(z)$, if $F(z) = f(g(z))$ provided that f is nonlinear and meromorphic and g is nonlinear and entire (g may be meromorphic when f is rational). F is said to be prime if every factorization of the above form implies that g is linear unless f is linear. In [2] Gross posed an open question whether there exists an entire periodic function which also is prime. Recently, Ozawa [3] exhibited such a function. His proof is complicated and is based on the fact that the order of the function constructed is finite (more precisely, the order is one). In this note we shall exhibit an infinite order periodic entire function which is prime. The proof of our example is relatively simpler than that of Ozawa.

Theorem. *There is an infinite order periodic entire function which is prime.*

Proof. Let $F(z) = (e^z - 1)\exp(\exp(-z + e^z))$. Clearly, F is entire, periodic, and of infinite order. We need to show that F is prime. In order to do so, we consider several cases and subcases.

Case 1. Suppose that $F(z) = f(g(z))$ where g is a polynomial of degree > 1 and f meromorphic. This case is ruled out by the asymmetrical behavior of F, since $F(z) \sim -e^{e^{-z}}$ in $\operatorname{Re} z < 0$, while $F(z)$ behaves more or less like $\exp(\exp e^z)$ on the positive real axis and its translation by $2n\pi i$, n integer.

Case 2. If $F = f(g(z))$, where g is meromorphic and f rational of degree p, say, then $T(r,F) \sim pT(r,g)$. Further, since $0, \infty$ have deficiency 1 for F it follows that any solution z_i of $f(z_i) = 0$ or ∞ has deficiency 1 for g. Thus, $f = 0$ and $f = 1$ have exactly 1 solution each. If $f(z_1) = 0$, the fact that zeros of $F = f \circ g$ are simple implies that z_1 is a simple zero of f. Hence f has degree 1, i.e., is bilinear.

Case 3. It remains to discuss the case $F(z) = f(g(z))$ when g is entire and transcendental and f is a transcendental meromorphic function (possibly entire).

From a result of Edrei [1] and the fact that the zeros of F, except for 0, are all imaginary, it follows that if f has an infinity of zeros, then g is a quadratic polynomial, which has been ruled out under Case 1. Thus, f has a finite set of zeros $w_1, \ldots, w_k$, $k \geqslant 1$.

Since $f(g(2n\pi i)) = 0$ there must be integers $n, m \neq 0$, such that $g(2m\pi i) = g((2n + 2m)\pi i)) =$ the same w_j. Since $f(g(z + 2n\pi i)) \equiv f(g(z))$ it follows that g has period $2n\pi i$. In the special case when $k = 1$ we see that $n = 1$.

Now if f has a pole at α and $g(z_0) = \alpha$, then z_0 is a pole of F. Thus, f has at most one such pole α and such an α, if it exists, is a value which g omits. We shall treat several subcases as follows:

Subcase I. Suppose f really has a pole α.

Then g omits α, ∞ and if γ is a zero of f, g takes γ only at the points $2m\pi i$ on the imaginary axis. By a theorem of Edrei [1] g has order 1 and so has the form

$$g = \alpha + Ae^{Bz}, \qquad A,B \text{ constants, } B \neq 0 .$$

Since g has period $2n\pi i$, Bn is an integer, say $k \neq 0$. Without loss of generality, we may take $\alpha = 0$, $A = 1$, so that

$$g(z) = e^{\frac{k}{n}z} ,$$

$$f(w) = (w^{n/k} - 1)\exp\left\{\frac{e^{w^{n/k}}}{w^{n/k}}\right\} .$$

Then f is certainly not meromorphic (it has no limit as $w \to 0$). Thus, Subcase I is impossible and we now consider only transcendental entire f.

Subcase II. f entire with at least two zeros, γ_1, γ_2; $f = p(z)e^{\alpha(z)}$, p polynomial, α entire and nonconstant.

Since f takes $\infty, \gamma_1, \gamma_2$ only on the imaginary axis, Edrei's theorem tells us that g has the order 1. From $f \circ g = F$ it follows that $p(g)$, which has order 1, must be of the form

$$p(g(z)) = (e^z - 1)e^{az+b}, \qquad a,b \text{ constants;}$$

and further that

$$(e^z - 1)e^{az+b}\, e^{\alpha(g)} = (e^z - 1)\exp(\exp(e^z - z)) .$$

Thus

$$az + b + \alpha(g) = e^{(e^z - z)} + 2k\pi i ,$$

for some integer k. Since g is periodic, we have a contradiction unless $a = 0$. Thus,

$$p(g(z)) = (e^z - 1)e^b .$$

Differentiating, $p'(g)g' = e^b e^z$, so g' has no zeros and is of order 1 and hence g has the form $\alpha + Ae^{Bz}$, which was shown to be impossible in the discussion of Subcase I.

Subcase III. Finally, suppose f is transcendental entire with exactly one zero. Without loss of generality, we may assume $f(0) = 0$:

$$f = z^n e^{\alpha(z)}$$

where α is entire and nonconstant.

In fact, $n = 1$, since the zeros of F are simple. Then

$$g(z) = (e^z - 1)e^{\beta(z)} , \qquad \beta \text{ entire,}$$

and

(1) $$\beta(z) + \alpha(g(z)) = 2k\pi i + \exp(e^z - z) .$$

We may suppose α has been so defined that $k = 0$.

Since g has period $2\pi i$ in this case (see remark preceding the discussion of subcase I) it follows that β has period $2\pi i$. The case of constant β would lead to $g = \text{const} + Ae^{Bz}$ which has already been shown to be impossible. Thus β is non-constant of period $2\pi i$.

Thus $\beta(z) = \Sigma_{-\infty}^{\infty} \lambda_n e^{nz}$, $\Sigma \lambda_n t^z$ convergent in $0 < |t| < \infty$, with at least one $\lambda_n \neq 0$, $n \neq 0$. β has type at least one. Suppose β is *not* of order 1, type 1. Then there exists $B > 1$ and a sequence $r_n \to \infty$ such that

$$M(\beta, r_n) > e^{Br_n} .$$

Choose a constant μ such that $\mu > 1$, $\mu^2 < B$. By Pólya's theorem [4] there is a constant $c > 0$ such that

$$M(e^{\beta}, \mu r_n) > \exp(cM(\beta, r_n)) > \exp(ce^{Br_n}) .$$

Moreover, the points z_n where this maximum is taken have $|\text{Re } z_n| \to \infty$, since otherwise $|e^{z_n}|$ would be bounded. Thus, $|e^{z_n} - 1| > \frac{1}{2}$ for large n and so

$$(2) \qquad M(g, \mu r_n) > \tfrac{1}{2} \exp(ce^{Br_n}) .$$

Using Pólya's theorem once more we see that

$$\begin{aligned} M(\alpha(g), \mu^2 r_n) &> M(\alpha, cM(g, \mu r_n)) \\ &> c' M(g, \mu r_n) \end{aligned}$$

for large n and some constant $c' > 0$. Comparing this with (1) and noting that the growth of β is negligible compared with that of g we find that from (1) $M(\alpha(g), \mu^2 r_n)$ must be comparable with $\exp(e^{\mu^2 r})$, while from (2) it is greater than $c'\exp(ce^{Br_n})$. Since $B > \mu^2$, this gives a contradiction.

Thus, β is of order 1, exponential type 1:

$$\beta = \lambda_{-1} e^{-z} + \lambda_0 + \lambda_1 e^z .$$

Further, $\lambda_{-1} \neq 0$, since otherwise β and g, F would be bounded in the left-half plane.

On $\text{Re } z = -R$ we have $\beta(-R + iy) = |\lambda_{-1}| e^R \cdot e^{-iy+y_0} + \lambda_0 + O(e^{-R})$, as $R \to \infty$; so as z describes $\text{Re } z = -R$, $\beta(z)$ describes an almost circular curve Γ around 0. An arc of this near $|\lambda_{-1}| e^R$ may be chosen which maps under $\beta \to e^{\beta}$ into a curve around 0 at radial distance roughly

$$e^{|\lambda_{-1}| e^R} \qquad (> e^{\frac{1}{2}|\lambda_{-1}| e^R} \text{ say}) .$$

g will behave similarly since $g \sim e^{\beta}$ on $\text{Re } z = -R$. On this last curve α will take at least one value $\alpha(z_0)$ such that $|\alpha(z_0)| > M(\alpha, \exp(\frac{1}{2}|\lambda_{-1}| e^R))$, so there will be a point on $\text{Re } z = -R$ where

$$|\alpha(g)| > M(\alpha, e^{\frac{1}{2}|\lambda_{-1}| e^R}) > ke^{\frac{1}{2}|\lambda_{-1}| e^R} ,$$

for some constant $k > 0$. But from (1), $\alpha(g) = \exp(e^z - z) - \beta(z)$ behaves like $O(e^{-z})$ in the left-half-plane. Thus, we have a contradiction and Subcase III is impossible. This also completes the proof of the theorem.

References

1. A. Edrei, Meromorphic Functions with Three Radially Distributed Values, *Trans. Amer. Math. Soc.* **78** (1955), 276–293.

2. F. Gross, Factorization of Entire Functions which are Periodic mod g, *Indian J. Pure and Applied Math.* **2** (1971), 561–571.

3. M. Ozawa, Factorization of Entire Functions, *Tohoku Math. Jour.* **27** (1975), 321–336.

4. G. Pólya, On an Integral Function of Integral Function, *Jour. London Math. Soc.* **1** (1926), 12–25.

Department of Mathematics
Imperial College
London, S.W. 7
England

Applied Mathematics Staff
Naval Research Laboratory
Washington, D.C. 20375

A UNIQUENESS THEOREM WITH APPLICATION TO THE ABEL SERIES

J. D. Buckholtz

1. Introduction. Let f be an entire function of exponential type less than 1. It is well known [1, p. 172] that if $f^{(n)}(n) = 0$, $n = 0,1,2,\ldots$, then $f \equiv 0$. The example

(1.1) $$f(z) = ze^{-z} = \sum_{k=0}^{\infty} k(-1)^{k-1}\frac{z^k}{k!},$$

which has exponential type 1 and satisfies $f^{(n)}(n) = 0$, $n = 0,1,2,\ldots$, shows that this result is, in a sense, best possible. The need for a sharper form of this result, however, in which exponential type is replaced by a more delicate growth measure, arises in the study of the Abel interpolation series. We prove such a theorem here and use it to determine necessary and sufficient conditions for an entire function f to be expandable in an Abel series.

Theorem 1.1. *Let f be an entire function such that $f^{(n)}(n) = 0$, $n = 0,1,2,\ldots$. If $f^{(k)}(0) = o(k) \to \infty$, then $f \equiv 0$. If f satisfies the weaker growth condition $f^{(k)}(0) = O(k)$, $k \to \infty$, then $f(z) = Cze^{-z}$ for some constant C.*

In view of the example (1.1), the second half of Theorem 1 implies the first.

Theorem 1.2. *An entire function f is the sum of its Abel interpolation series,*

(1.2) $$f(z) = \sum_{n=0}^{\infty} f^{(n)}(n)\frac{z(z-n)^{n-1}}{n!},$$

if and only if f satisfies the conditions

(i) $f^{(k)}(0) = o(k)$, $k \to \infty$, *and*

(ii) $\sum_{n=1}^{\infty} f^{(n)}(n)\frac{(-n)^{n-1}}{n!}$ *converges.*

An easy argument allows one to obtain Theorem 1.2 from Theorem 1.1; we give it here. Condition (ii) is known [3, Lemma 2] to be both necessary and sufficient for the uniform convergence on compact sets of the series

(1.3) $$S(z) = \sum_{n=0}^{\infty} f^{(n)}(n)\frac{z(z-n)^{n-1}}{n!},$$

whose sum need not be $f(z)$. It was proved in [3] that (i) is a necessary condition for (1.2) to hold. We need only show then that (i) and (ii) imply (1.2). Condition (ii) implies that the function S defined by (1.3) is entire, and from successive differentiation we obtain $S^{(n)}(n) = f^{(n)}(n)$, $n = 0,1,2,\ldots$. Therefore, the entire function $g(z) = f(z) - S(z)$ satisfies $g^{(n)}(n) = 0$, $n = 0,1,2,\ldots$. Since

$$S(z) = \sum_{n=0}^{\infty} S^{(n)}(n) \frac{z(z-n)^{n-1}}{n!} ,$$

the necessity of (i) implies the estimate $S^{(k)}(0) = o(k)$. Therefore, $g^{(k)}(0) = f^{(k)}(0) - S^{(k)}(0) = o(k)$. Theorem 1.1 allows us to conclude that $g \equiv 0$, and this completes the proof.

2. Kernel Approximation. We use the Polya representation to obtain a weaker form of Theorem 1.1. The proof in this case depends on showing that, for fixed z, the Polya kernel function e^{zw} can be uniformly approximated on the circle $|w| = 1$ by polynomials in the variable $u = we^w$. This avoids certain difficulties inherent in the kernel expansion and summability techniques employed by Buck [2].

Lemma 2.1. *Let* h *be an entire function such that* $h^{(n)}(n) = 0$, $n = 0,1,2,\ldots$. *If* $h^{(k)}(0) = O(k^{-2})$, *then* $h \equiv 0$.

Proof. The Borel transform

$$H(w) = \sum_{k=0}^{\infty} \frac{h^{(k)}(0)}{w^{k+1}}$$

of h is analytic for $|w| > 1$ and continuous for $|w| \geqslant 1$. Therefore, we can take the circle $|w| = 1$ for the path of integration in the Polya representation and write

$$h(z) = \frac{1}{2\pi i} \int_{|w|=1} e^{zw} H(w)\, dw .$$

Differentiation under the integral sign yields

$$h^{(n)}(n) = \frac{1}{2\pi i} \int_{|w|=1} (we^w)^n H(w)\, dw , \qquad n = 0,1,2,\ldots .$$

The mapping $u = we^w$ is univalent and starlike for $|w| < 1$. Let $w = \varphi(u)$ denote the inverse of the restriction of $u = we^w$ to $|w| \leqslant 1$. Then the domain of φ is a compact set K whose complement is connected; the function φ is continuous on K and analytic on the interior of K.

Let z be fixed, set

$$M = \max_{|w|=1} |H(w)| ,$$

and suppose $\epsilon > 0$. Mergelyan's Theorem [7, p. 423] allows us to choose a polynomial $P(u) = \Sigma_{t=0}^{N} c_t u^t$ such that

$$|e^{z\varphi(u)} - P(u)| < \frac{\epsilon}{M}$$

for all u in K. This implies that $|e^{zw} - P(we^w)| < \epsilon/M$ for $|w| \leqslant 1$. Since

$$\frac{1}{2\pi i} \int_{|w|=1} P(we^w) H(w)\, dw = \sum_{t=0}^{N} \frac{c_t}{2\pi i} \int_{|w|=1} (we^w)^t H(w)\, dw$$

$$= \sum_{t=0}^{N} c_t h^{(t)}(t) = 0 ,$$

we have

$$h(z) = \frac{1}{2\pi i}\int_{|w|=1} \{e^{zw} - P(we^{w})\}H(w)\,dw .$$

Therefore, $|h(z)| \leqslant M(\epsilon/M) = \epsilon$. Since ϵ and z are arbitrary, it follows that $h \equiv 0$.

3. A Reduction Operator. The rest of the proof of Theorem 1.1 depends on the operator B defined on the family F of entire functions which vanish at 0 by

$$(Bf)(z) = \frac{f(z)}{z} - f'(0) + \int_0^z \frac{f(t)}{t}\,dt .$$

The equation $Bf = g$ can be solved explicitly; we obtain

$$f(z) = ze^{-z}\int_0^z e^{t}g'(t)\,dt + Cze^{-z} ,$$

where C is an arbitrary constant. This allows us to obtain the general solution of $B^jf = g$ for each positive integer j. In particular, the general solution of $B^3f = 0$ is given by

(3.1) $$f(z) = C_1ze^{-z} + C_2\left(z^2 - \frac{z^3}{2}\right)e^{-z} + C_3\left(z^3 - \frac{5z^4}{6} + \frac{z^5}{8}\right)e^{-z} .$$

It is not hard to verify that B has the properties

(3.2) $$(Bf)^{(k)}(0) = \frac{f^{(k)}(0)}{k} + \frac{f^{(k+1)}(0)}{k+1} , \qquad k = 1,2,3,\ldots ,$$

and

(3.3) $$(Bf)^{(n)}(n) = \frac{f^{(n)}(n)}{n} , \qquad n = 1,2,3,\ldots .$$

If f satisfies a growth condition of the form $f^{(k)}(0) = O(k^t)$, then it follows from (3.2) that Bf satisfies the growth condition $(Bf)^{(k)}(0) = O(k^{t-1})$. If $f^{(n)}(n) = 0$, $n = 0,1,2,\ldots$, then (since $(Bf)(0) = 0$) equation (3.3) implies that $(Bf)^{(n)}(n) = 0$, $n = 0,1,2,\ldots$.

Let f be an entire function which satisfies the hypotheses of the second half of Theorem 1.1, apply B three times, and set $h = B^3f$. Then $h^{(k)}(0) = O(k^{-2})$ and $h^{(n)}(n) = 0$, $n = 0,1,2,\ldots$. Therefore, $B^3f = h \equiv 0$ by Lemma 2.1, and f must be a function of the form (3.1). To complete the proof, note that the growth condition $f^{(k)}(0) = O(k)$ forces $C_2 = C_3 = 0$.

4. Extensions. The techniques of §2 and §3 can be used to study a variety of problems involving sequences of linear functionals. The appropriate setting is to regard the linear functional $L_n(f) = f^{(n)}(n)$ as arising from the infinite order differential operator De^D in the following way:

$$L_n(f) = \{(De^D)^n f\}(0) , \qquad n = 0,1,2,\ldots .$$

Theorem 1.1 can then be regarded as a consequence of geometric properties of the mapping $u = we^w$. This point of view has been used extensively by Gelfond [6] and others to study sequences of linear functionals of the form

$$L_n(f) = \{[G(D)]^n f\}(0), \qquad n = 0,1,2,\ldots, \tag{4.1}$$

where the function G is analytic and univalent in a neighborhood of 0. Some appreciation of the power of this point of view can be obtained from the following result, due to DeMar [5] and Gelfond [6]:

Suppose G is analytic on a simply connected region Ω containing 0. Let f be an entire function of exponential type subject to the following growth condition: the Borel transform of f is regular on the complement of Ω. Then the condition $L_n(f) = 0$, $n = 0,1,2,\ldots$, is sufficient to imply that $f \equiv 0$ *if and only if* G is univalent on Ω.

The necessity of the univalence of G is easy to establish. If $G(\alpha) = G(\beta)$ for points α, β in Ω, then the function

$$f_1(z) = e^{\alpha z} - e^{\beta z}$$

satisfies $([G(D)]^n f_1)(z) = [G(\alpha)]^n e^{\alpha z} - [G(\beta)]^n e^{\beta z}$ so that $L_n(f_1) = [G(\alpha)]^n - [G(\beta)]^n = 0$. Similarly, if $G'(\alpha) = 0$ for α in Ω, the function

$$f_2(z) = ze^{\alpha z}$$

satisfies $L_n(f_2) = n[G(\alpha)]^{n-1}G'(\alpha) = 0$.

The result quoted at the beginning of §1 corresponds to the special case $G(w) = G_1(w) = we^w$ and $\Omega = \{|w| < 1\}$. The function $f_2(z) = ze^{-z}$ arises from the zero of $G_1'(w) = (w+1)e^w$ at $w = -1$.

The arguments used in §2 depend on the following three properties of the mapping function $G_1(w) = we^w$:

(i) G_1 is univalent in the disk $|w| < 1$;

(ii) G_1 is analytic on the closed disk $|w| \leq 1$; and

(iii) the image of the circle $|w| = 1$ under G_1 is a simple closed curve.

Essentially identical results can be obtained for any function G satisfying (i), (ii) and (iii). In the general case the disk $|w| < 1$ is replaced by $|w| < r$, where r denotes the radius of univalence of G; the growth conditions are modified by changing $O(k)$ and $o(k)$ to $O(kr^k)$ and $o(kr^k)$.

Finding a suitable reduction operator in the general setting is difficult. The operator B of §3 bears no apparent relation to the mapping G_1, yet its existence certainly justifies a search for something equivalent in the general case. Fortunately, such an operator does exist: it is given by

$$(Rf)(z) = \int_0^z \left\{\frac{DG'(D)}{G(D)}\hat{f}\right\}(t)\,dt,$$

where $\hat{f}$ is the function defined by $\hat{f}(z) = f(z)/z$. To see that R reduces to B when $G = G_1$, note that

$$\frac{wG_1'(w)}{G_1(w)} = \frac{w(w+1)e^w}{we^w} = w+1,$$

and that

$$\int_0^z \{(D+1)\hat{f}\}(t)\,dt = [\hat{f}(t)]_0^z + \int_0^z \hat{f}(t)\,dt$$

$$= \frac{f(z)}{z} - f'(0) + \int_0^z \frac{f(t)}{t}\,dt \ .$$

The domain of R is restricted to entire functions which vanish at 0 and whose exponential type does not exceed the radius of analyticity of $wG'(w)/G(w)$. This is necessary to insure that $\hat{f}$ belongs to the domain of the infinite order differential operator $DG'(D)/G(D)$. Results corresponding to those of §3 can still be obtained, but the difficulties are significantly greater.

A treatment of the general case will appear elsewhere [4]. We confine ourselves here to a discussion of the special case in which the functionals L_n are central differences. In addition to being fairly typical, this case is of particular interest in that it can be used to obtain a new proof of the following theorem of Polya [1]:

Theorem 4.1. *If an entire function* f *satisfies* $f(n) = 0,\ n = 0,\pm1,\pm2,\dots$ *and* $f^{(k)}(0) = o(\pi^k)$, *then* $f \equiv 0$. *More generally, if* f *vanishes at the integers and* $f^{(k)}(0) = O(\pi^k)$, *then* $f(z) = C \sin \pi z$ *for some constant* C.

(Polya did in fact prove considerably more than this. For the purpose of illustrating our technique, the weaker version stated above is adequate.)

The choice $G(w) = G_0(w) = 2 \sin h(w/2)$ yields

$$L_n(f) = \Delta^n f\left(\frac{-n}{2}\right) = (-1)^n \sum_{k=0}^{n} \binom{n}{k} (-1)^k f\left(\frac{-n}{2} + k\right), \qquad n = 0,1,2,\dots,$$

the central differences of f. The function G_0 has π for its radius of univalence and the image of $|w| = \pi$ under G_0 is a simple closed curve with two cusps, corresponding to the zeros of G_0' at $w = \pm\pi i$. These zeros give rise to the functions $ze^{\pi iz}$ and $ze^{-\pi iz}$, all of whose central differences vanish. The same is true for linear combinations of these functions, and therefore for the functions $z \cos \pi z$ and $z \sin \pi z$. Corresponding to Theorem 1.1 we have

Theorem 4.2. *Let* f *be an entire function such that* $\Delta^n f(-n/2) = 0,\ n = 0,1,2,\dots$. *If* $f^{(k)}(0) = o(k\pi^k)$, *then* $f \equiv 0$. *If* f *satisfies the weaker growth condition* $f^{(k)}(0) = O(k\pi^k)$, *then* f *is a linear combination of* $z \cos \pi z$ *and* $z \sin \pi z$.

To obtain Theorem 4.1 from Theorem 4.2, let f be an entire function satisfying the growth condition $f^{(k)}(0) = O(\pi^k)$ and such that $f(n) = 0,\ n = 0,\pm1,\pm2,\dots$. Write $f = f_e + f_o$, where f_e and f_o are the even and odd parts of f. Then both f_e and f_o satisfy the same growth condition as f, and both vanish at the integers. Further, the central differences of f_e vanish: the central differences of even order because f_e vanishes at the integers, and the central differences of odd order because f_e is an even function. The growth condition $f_e^{(k)}(0) = O(\pi^k)$ implies $f_e^{(k)}(0) = o(k\pi^k)$ and therefore that $f_e \equiv 0$ by Theorem 4.2. This establishes that f is odd, and consequently that the function $F(z) = zf(z)$ is even. The growth condition on f implies the growth condition $F^{(k)}(0) = O(k\pi^k)$ for F. Now repeat the previous argument with f_e replaced by F. Theorem 4.2 implies that F is a linear combination of the functions $z \cos \pi z$ and $z \sin \pi z$. Since F is even we must have $F(z) = Cz \sin \pi z$ for some constant C; therefore $f(z) = C \sin \pi z$.

References

1. R. P. Boas, Jr., **Entire Functions**, Academic Press, New York, 1954.

2. R. C. Buck, Interpolation series, *Trans. Amer. Math. Soc.* **64** (1948), 283–298.

3. J. D. Buckholtz, Series expansions of analytic functions, *J. Math. Anal. Appl.* **41** (1973), 673–684.

4. J. D. Buckholtz, A geometric criterion for uniqueness of linear functionals on entire functions, to appear in *Ill. J. Math.*

5. R. F. DeMar, A uniqueness theorem for entire functions, *Proc. Amer. Math. Soc.* **16** (1965), 69–71.

6. A. O. Gelfond, **Differenzenrechnung**, Deutscher Verlag der Wissenschaften, Berlin, 1958.

7. W. Rudin, **Real and Complex Analysis**, 2nd ed., McGraw-Hill, New York, 1974.

University of Kentucky
Lexington, Kentucky 40506

ON THE DISTRIBUTION OF VALUES OF MEROMORPHIC FUNCTIONS OF SLOW GROWTH

Meledath Damodaran

I. Introduction. In this note we consider the following problem:

If $f(z)$ *is a meromorphic function of order* ρ $(\leqslant\infty)$, *determine for how large a set* A *can*

$$\lim_{r\to\infty} m(r,a) = \infty \qquad \text{for every } a \in A. \tag{1.1}$$

For entire functions this question is treated in [3]. We refer the reader to this reference for background and discussion in depth. Briefly, the situation in that case can be described as follows:

If $\rho > \frac{1}{2}$, *a necessary and sufficient condition for there to exist an entire function satisfying (1.1) is that* A *be of capacity zero; if* $\rho = \frac{1}{2}$, *it is necessary and sufficient that* A *be contained in an* F_σ *set of capacity zero; if* $\rho < \frac{1}{2}$, *then* $A = \{\infty\}$.

Since it is classical that (1.1) always implies A has capacity zero, the general case of meromorphic functions and $\rho > \frac{1}{2}$ is then covered by the above mentioned results for entire functions.

For meromorphic functions, we have found that the order of the function imposes no additional restrictions on A.

Theorem 1. *Let* A *be an arbitrary set of capacity zero. Then we have the following:*

a) *Given* $\lambda(r)$ *tending to infinity as* $r\to\infty$, *there exists a meromorphic function* f *of order zero, which satisfies (1.1) and such that*

$$T(r,f) = O(\lambda(r)(\log r)^3) \qquad (r\to\infty) \tag{1.2}$$

b) *Given* ρ, $0\leqslant\rho\leqslant\infty$, *there exists a meromorphic function* f *of order* ρ, *which satisfies (1.1).*

Our reason for distinguishing statement a) in Theorem 1 is that it turns out that very slow growth entails severe restrictions on A.

Theorem 2. *If* $f(z)$ *is a meromorphic function satisfying*

$$T(r,f) = O((\log r)^2) \qquad (r\to\infty) \tag{1.3}$$

then the set A *of (1.1) can have at most one element.*

The author wishes to thank Professor Hayman for pointing out that Theorem 2 is a direct consequence of an earlier theorem of P. D. Barry [1, Theorem 5].

II. Proof of Theorem 1. As we have mentioned, all values of ρ in the interval $\frac{1}{2} < \rho \leq \infty$ are covered in [3]. In fact, what we require here is a suitable version of [3, Lemma 2] for the growth rates $0 < \rho \leq \frac{1}{2}$. Once this is achieved (in the Lemma below) only simple modifications of [3, §7] are needed to complete the proof. Because of the routine nature of the necessary changes we omit the duplication of the arguments of [3, §7].

As in [3, Lemma 2] we construct a function which approximates values $\zeta_{k,n}$ on subsets $E_{k,n}$ of $\{2^n \leq |z| \leq 2^{n+1}, |\arg z| \leq \pi/4\}$, where

(2.1)
$$E_{k,n} = \begin{cases} \{z: 2^n \leq |z| \leq 2^{n+1},\ (2k-1)\frac{\pi}{8\nu_n} \leq \arg z \leq 2k\frac{\pi}{8\nu_n}\} & (n \text{ even}) \\ \{z: 2^n \leq |z| \leq 2^{n+1},\ (2k-1)\frac{\pi}{8\nu_n} \leq -\arg z \leq 2k\frac{\pi}{8\nu_n}\} & (n \text{ odd}) \end{cases}$$

for $n = 1,2,\ldots,\ k = 1,2,\ldots,\nu_n$, where $\{\nu_n\}$ is a sequence of integers tending to ∞. With these notations we have the following.

Lemma. *Let* $\zeta_{k,n}$ *be a sequence of complex numbers satisfying*

(2.2)
$$|\zeta_{k,n}| \leq 2^n, \quad n \geq n_o, \quad k = 1,2,\ldots,\nu_n .$$

Then

a) *given* $\lambda(r)$ *tending to* ∞, *there exists a sequence of integers* $\{\nu_n\}$ *such that* $\nu_n \to \infty$ *as* $n \to \infty$, *and a meromorphic function* f *satisfying (1.2) and such that*

(2.3)
$$|f(z) - \zeta_{k,n}| < \frac{K_o}{2^n} \qquad (z \in E_{k,n})$$

where $K_o > 0$ *is a constant;*

b) *given* ρ, $0 < \rho \leq \frac{1}{2}$, *there exists* $\{\nu_n\}$ *as in* a) *and a meromorphic function* f *of order* ρ *satisfying (2.3).*

Proof of Lemma. From the proof of [3, Lemma 2] we obtain a function $h(z)$ analytic in $\{z \in C, |\arg z| \leq \frac{\pi}{2}\}$ and such that

(2.4)
$$\left| \frac{h(z)}{z+1} - \zeta_{k,n} \right| < \frac{K_1}{|z+1|} \qquad (z \in E_{k,n})$$

and for $2^n \leq r \leq 2^{n+1}$

(2.5)
$$\log M(r) \leq 16(\log \nu_n + n)e^{\frac{10^4}{\pi}\nu_n} \qquad (n > n_o)$$

where $M(r)$ stands for

$$\max_{\substack{|z|=r \\ \operatorname{Re} z > 0}} |h(z)| .$$

To prove part a) of the lemma we assume, without loss of generality, that $\lambda(r)$ is monotone increasing and choose ν_n to tend to ∞ slowly enough so that

$$(2.6)\qquad 16(\log \nu_n + n)e^{\frac{10^4}{\pi}\nu_n} < (\log 2^n)\lambda(2^{n-1}) \qquad (n > n_0)\,.$$

Then from (2.5) and (2.6) we have

$$(2.7)\qquad \log M(r) < \lambda\left(\frac{r}{2}\right)\log r \qquad (r > r_0)\,.$$

Applying [5, Theorem 1] we obtain, for $\eta \in (0,1)$ and $q = 1 + \delta$ where $0 < \delta < \sin(\pi/8)$, a function $F(z)$ meromorphic in C such that

$$(2.8)\qquad |h(z) - F(z)| \leqslant \eta \qquad (-\tfrac{\pi}{4} \leqslant \arg z \leqslant \tfrac{\pi}{4}\,,\ |z| > 1)$$

and

$$(2.9)\qquad T(r,F) \leqslant A\Bigg[\int_1^r\!\!\int_1^t \frac{\log M(q\tau)}{\tau t}\,d\tau\,dt + \int_1^r \frac{\log M(qt)}{t}\,dt$$

$$+ \log^+ M(q^{\ell+1}r) + (\log r)^3 + 1\Bigg] \qquad (r \geqslant r_0)$$

where ℓ is the least integer such that $q^{\ell-1} > (1 - \sin(\pi/8))^{-1}$. By virtue of (2.7),

$$\int_1^r\!\!\int_1^t \frac{\log M(q\tau)}{t\tau}\,d\tau\,dt$$

$$\leqslant \lambda(r)\int_1^r\!\!\int_1^t \frac{\log(q\tau)}{\tau t}\,d\tau\,dt$$

$$= \lambda(r)\int_1^r \frac{1}{t}\left(\frac{(\log(qt))^2}{2} - \frac{(\log q)^2}{2}\right)dt$$

$$= \lambda(r)\left[\frac{(\log(qr))^3}{6} - \frac{(\log q)^2}{2}\log r - \frac{(\log q)^3}{6}\right]$$

$$\leqslant B_1\lambda(r)(\log r)^3\,, \qquad \text{for some constant } B_1 > 0\,.$$

A similar estimate holds for the second term in the square bracket in (2.9), and hence we get

$$(2.10)\qquad T(r,F) \leqslant B\lambda(r)(\log r)^3\,.$$

If we now let

$$(2.11)\qquad f(z) = \frac{F(z)}{z+1}\,,$$

from (2.4), (2.8), (2.10) and (2.11) it follows that f satisfies (2.3) and (1.2), and the proof of part a) is complete.

The proof of part b) is along the same lines as that of a) and we indicate the necessary modifications. Here, replace $h(z)/(z+1)$ in (2.4) by $h(z)$ and apply [5, Theorem 2] to obtain a meromorphic function f in C having the following three properties:

$$(2.12)\qquad |h(z) - f(z)| \leqslant e^{-|z|^\rho} \qquad (|\arg z| \leqslant \tfrac{\pi}{4}\,,\ r > r_0)\,;$$

(2.13) $$|f(z)| \leqslant e^{-|z|^\rho} \qquad (|\arg z - \theta_0| \leqslant \frac{\alpha}{4},\ r > r_0)$$

where θ_0 and α are such that $\rho < 2\pi/\alpha$ and

$$\{z \neq 0,\ |\arg z - \theta_0| \leqslant \frac{\alpha}{2}\} \cap \{z \neq 0,\ |\arg z| \leqslant \frac{\pi}{2}\} = \emptyset ;$$

(2.14) $$T(r,f) \leqslant Ar^\rho, \text{ for some constant } A.$$

Now, for $z \in E_{k,n}$

$$\begin{aligned} |f(z) - \zeta_{k,n}| &\leqslant |f(z) - h(z)| + |h(z) - \zeta_{k,n}| \\ &\leqslant e^{-|z|^\rho} + \frac{K_1}{|z|} \\ &\leqslant \frac{K_0}{2^n} \end{aligned}$$

so f satisfies (2.3).

From (2.13) it follows that

(2.15) $$\begin{aligned} m(r,\frac{1}{f}) &= \frac{1}{2\pi}\int_0^{2\pi} \log^+ \frac{1}{|f(re^{i\theta})|}\, d\theta \\ &\geqslant \frac{1}{2\pi}\int_{|\arg z - \theta_0| \leqslant \frac{\alpha}{4}} |z|^\rho\, d\theta \\ &\geqslant K_2 r^\rho \end{aligned}$$

so (2.14) and (2.15) together show that f has order ρ. The proof of b) is now complete.

III. **Proof of Theorem 2.** Suppose A has more than one element. Without loss of generality we may then assume $0,\infty \in A$, and by elementary properties of canonical products of genus 0 (cf. [4, p. 21])

$$f(z) = cz^k \prod_{n=1}^{\infty} \frac{1 - \frac{z}{a_n}}{1 - \frac{z}{b_n}} = \frac{g(z)}{h(z)}$$

where $g(z)$, $h(z)$ are entire and satisfy ([2, p. 47])

(3.1) $$\begin{aligned} &\log M(r,g) = O((\log r)^2) && (r \to \infty) \\ &\log M(r,h) = O((\log r)^2) && (r \to \infty) . \end{aligned}$$

From a theorem of Barry [1, Theorem 5], it follows that the conditions (3.1) imply the existence of a sequence $r_m \to \infty$ and constants k_1 and k_2 such that for all $z = r_m e^{i\theta}$,

$$k_1 \frac{M(r_m,g)}{M(r_m,h)} \leqslant |f(z)| \leqslant k_2 \frac{M(r_m,g)}{M(r_m,h)} .$$

Theorem 2 then follows.

References

1. P. D. Barry, The minimum modulus of integral functions of small order, *Bull. A.M.S.* **67** (1961), 231–234.

2. R. Boas, **Entire functions**, Academic Press, New York, 1954.

3. D. Drasin and A. Weitsman, The growth of the Nevanlinna proximity function and the logarithmic potential, *Indiana Univ. Math. Jour.* **20** (1971), 699–715.

4. W. Hayman, **Meromorphic functions**, Clarendon Press, Oxford, 1964.

5. L. Ter-Israjelian, Uniform and tangential approximation of functions holomorphic in an angle by meromorphic functions with estimates on their growth (in Russian), *Izv. Akad. Nauk Arm. SSR Matem.* 6 (1971), 67–79.

Purdue University
West Lafayette, Indiana 47907

SUBORDINATION

Peter Duren

The purpose of this brief survey is to gather selected information related to the concept of subordination, to describe some old results which deserve to be more widely known, and to point out some unsolved problems. A more complete discussion will appear in my forthcoming book [10] on univalent functions.

Let f and g be analytic in the unit disk, with $f(0) = g(0)$. Suppose also that f is univalent, and that its range contains that of g. Then the Schwarz lemma easily shows $g(z) = f(\omega(z))$, where ω is analytic and $|\omega(z)| \leqslant |z|$. In general, g is said to be *subordinate* to f (written $g \prec f$) if the two functions are related in this way. This definition does not require that f be univalent.

Mean domination.

If $g \prec f$, it is clear that the maximum modulus of f dominates that of g: $M_\infty(r,g) \leqslant M_\infty(r,f)$ for $0 \leqslant r < 1$. J. E. Littlewood [19] extended this to integral means of order p: $M_p(r,g) \leqslant M_p(r,f)$ for $0 < p < \infty$ and $0 \leqslant r < 1$, where

$$M_p(r,f) = \left\{ \int_0^{2\pi} |f(re^{i\theta})|^p \, d\theta \right\}^{1/p} .$$

The elegant proof by F. Riesz, generalizing the theorem to subharmonic functions, may be found in [9].

Suppose now that $f(0) = g(0) = 0$ and let

$$f(z) = \sum_{n=1}^{\infty} a_n z^n \; ; \qquad\qquad g(z) = \sum_{n=1}^{\infty} b_n z^n .$$

The subordination condition does not imply $|b_n| \leqslant |a_n|$; for example, $z^2 \prec z$. Nevertheless, W. Rogosinski [27] observed that $g \prec f$ does imply

$$\sum_{k=1}^{n} |b_k|^2 \leqslant \sum_{k=1}^{n} |a_k|^2 , \qquad n = 1,2,\ldots .$$

The proof is based on Littlewood's theorem. As a simple corollary, if f has bounded coefficients, then $b_n = O(\sqrt{n})$. Surprisingly, this estimate is best possible: Rogosinski showed that b_n need not be $o(\sqrt{n})$.

G. M. Goluzin [14] extended Rogosinski's theorem by showing that if $g \prec f$ and $\lambda_1 \geqslant \lambda_2 \geqslant \cdots \geqslant 0$, then

$$\sum_{n=1}^{\infty} \lambda_n |b_n|^2 \leqslant \sum_{n=1}^{\infty} \lambda_n |a_n|^2 .$$

The proof merely uses summation by parts to reduce the inequality to Rogosinski's case, but there are two interesting corollaries. Let

$$A(r,f) = \iint_{|z| \leqslant r} |f'(\rho e^{i\theta})|^2 \rho \, d\rho \, d\theta = \pi \sum_{n=1}^{\infty} n |a_n|^2 r^{2n}$$

denote the area of the (multisheeted) image of the disk $|z| \leqslant r$ under the mapping f. Goluzin's result implies that if $g \prec f$, then

$$A(r,g) \leqslant A(r,f) , \qquad 0 \leqslant r \leqslant \frac{1}{\sqrt{2}} ,$$

and

$$M_2(r,g') \leqslant M_2(r,f') , \qquad 0 \leqslant r \leqslant \tfrac{1}{2} .$$

In both cases, the example $z^2 \prec z$ shows the bounds on r are best possible.

Generalized Bieberbach conjecture.

Up to this point we have made no univalence assumptions. Suppose now that f belongs to the class S of univalent functions in the disk with $f(0) = 0$ and $f'(0) = 1$. The *generalized Bieberbach conjecture* asserts that $|b_n| \leqslant n$ if $g \prec f$ for some $f \in S$. This follows from the Schwarz lemma for $n = 1$, and Littlewood [19] proved it for $n = 2$. It is true for all n under the additional assumption that f is starlike [27], typically real [27], or close-to-convex [22].

The general estimate $|b_n| < \frac{e}{2} n$ can be obtained by combining Littlewood's theorem with the recent result of A. Baernstein [1] that $M_1(r,f) \leqslant M_1(r,k)$, $0 \leqslant r < 1$, for all $f \in S$, where $k(z) = z(1-z)^{-2}$ is the Koebe function. The constant e/2 seems to be the best known.

M. S. Robertson [24] recently observed that the generalized Bieberbach conjecture is a consequence of *Robertson's conjecture* [21] on the coefficients of odd univalent functions. The most general odd function in S has the form

$$h(z) = \sqrt{f(z^2)} = z + c_3 z^3 + c_5 z^5 + \cdots$$

for some $f \in S$. Comparing coefficients of z^{2n} in the equation $[h(z)]^2 = f(z^2)$, we find

$$a_n = c_1 c_{2n-1} + c_3 c_{2n-3} + \cdots + c_{2n-1} c_1 , \qquad c_1 = 1 .$$

Thus, the Cauchy-Schwarz inequality gives

$$|a_n| \leqslant \sum_{k=1}^{n} |c_{2k-1}|^2 , \qquad n = 1,2,\ldots .$$

Robertson's conjecture asserts that this last sum is $\leqslant n$. This would imply not only the Bieberbach conjecture, but also the generalized Bieberbach conjecture, as the following argument reveals.

Suppose $g \prec f$, and write $g(z) = f(\omega(z))$. Then

$$\varphi(z) = \frac{h(\sqrt{z})}{\sqrt{z}} = 1 + c_3 z + c_5 z^2 + \cdots$$

and $[\varphi(z)]^2 = f(z)/z$. Thus

$$g(z) = \omega(z)\{1 + c_3\omega(z) + c_5[\omega(z)]^2 + \cdots\}^2 .$$

Let

$$s_n(z) = \sum_{k=1}^{n} c_{2k-1} z^{k-1}$$

denote the nth partial sum of φ. Then since $\omega(0) = 0$,

$$b_n = \frac{1}{2\pi i} \int_{|z|=r} \frac{\omega(z)[s_n(\omega(z))]^2}{z^{n+1}} dz .$$

Since $s_n(\omega(z)) \prec s_n(z)$, it follows from Littlewood's theorem that

$$\begin{aligned} |b_n| &\leq r^{-n}[M_2(r, s_n \circ \omega)]^2 \\ &\leq r^{-n}[M_2(r, s_n)]^2 \\ &= r^{-n} \sum_{k=1}^{n} |c_{2k-1}|^2 r^{2k-2} . \end{aligned}$$

Let $r \to 1$ to conclude that

$$|b_n| \leq \sum_{k=1}^{n} |c_{2k-1}|^2 .$$

Because Robertson [21] proved Robertson's conjecture for $n = 3$ and S. Friedland [11] verified it for $n = 4$, this establishes the generalized Bieberbach conjecture up to $n = 4$. For larger n, both conjectures remain unsettled.

Sharpened forms of the Schwarz lemma.

The Schwarz lemma asserts that if $\omega(z)$ is analytic in $|z| < 1$, $|\omega(z)| \leq 1$, and $\omega(0) = 0$, then $|\omega(z)| \leq |z|$ and $|\omega'(0)| \leq 1$. The first conclusion can be sharpened by describing the exact region of variability of $\omega(z_0)$ for fixed z_0, under the normalizing assumption $\omega'(0) \geq 0$. Let Δ_{z_0} denote the closed region containing the disk $|w| \leq |z_0|^2$ and bounded by an arc of the circle $|w| = |z_0|^2$ and by the two circular arcs tangent to this circle at $\pm i\,|z_0|z_0$ and passing through z_0. (See diagram.)

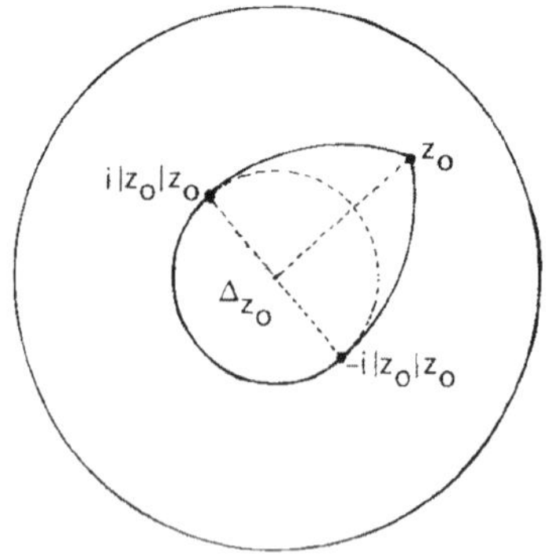

The following result is due to Rogosinski [25].

Rogosinski's lemma. *Suppose* $0 < |z_0| < 1$. *Then for all functions* $\omega(z)$ *analytic and satisfying* $|\omega(z)| < 1$ *in* $|z| < 1$, *with* $\omega(0) = 0$ *and* $\omega'(0) \geq 0$, *the region of values of* $\omega(z_0)$ *is precisely* Δ_{z_0}.

The proof uses the invariant form of the Schwarz lemma to place $\omega(z_0)$ in the union of a certain family of disks centered on the segment from 0 to z_0, with $t = \omega'(0)$ acting as a parameter, $0 \leq t \leq 1$. Functions of the form

$$\omega(z) = z\,\frac{\alpha z + t}{1 + t\alpha z}\,, \qquad |\alpha| < 1, \qquad 0 \leq t \leq 1\,,$$

show that the entire region Δ_{z_0} is covered. See Goluzin's book [16] for further details.

A similar argument allows an extension of the Schwarz lemma to describe the region of variability of the derivative $\omega'(z_0)$. This result is essentially due to J. Dieudonné [8]. A proof may be found in Carathéodory's book [7, p. 19].

Dieudonné's lemma. *Suppose* $0 < |z_0| < 1$ *and* $|w_0| < 1$. *Then for all functions* $\omega(z)$ *analytic and satisfying* $|\omega(z)| < 1$ *in* $|z| < 1$, *with* $\omega(0) = 0$ *and* $\omega(z_0) = w_0$, *the region of values of* $\omega'(z_0)$ *is precisely the disk*

$$\left| w - \frac{w_0}{z_0} \right| \leq \frac{|z_0|^2 - |w_0|^2}{|z_0|(1 - |z_0|^2)}\,.$$

Corollary 1. *If* $\omega(z)$ *is analytic in* $|z| < 1$, $|\omega(z)| < 1$, *and* $\omega(0) = 0$, *then*

$$|\omega'(z)| \leq \begin{cases} 1\,, & |z| \leq \sqrt{2} - 1 \\ \dfrac{(1 + |z|^2)^2}{4|z|(1 - |z|^2)}\,, & |z| > \sqrt{2} - 1\,. \end{cases}$$

The bound is sharp for each z.

Corollary 2. *If* $g \prec f$ *and* $0 < p < \infty$, *then* $M_p(r,g') \leq M_p(r,f')$ *for* $r \leq \sqrt{2} - 1$.

The proof of Corollary 2 goes as follows. Write $g(z) = f(\omega(z))$ and note that

$$|g'(z)| = |f'(\omega(z))|\,|\omega'(z)| \leq |f'(\omega(z))| \qquad \text{for } |z| \leq \sqrt{2} - 1\,,$$

by Corollary 1. But $f'(\omega(z)) \prec f'(z)$, so Littlewood's theorem gives the result.

It is apparently an open problem to find the sharp upper bound for r in Corollary 2. Goluzin's theorem (stated above) increases it to ½ for $p = 2$, and the example $z^2 \prec z$ again shows nothing better is possible for any p. For $p = \infty$, the sharpness of Dieudonné's lemma implies that $\sqrt{2} - 1$ is best possible. Simply choose $f(z) = z$.

Pointwise domination.

According to Littlewood's theorem, the subordination condition $g \prec f$ implies that f dominates g in the mean. We now turn to one of the deepest results in subordination theory, asserting that under an appropriate normalization, $|f(z)|$ actually dominates $|g(z)|$ *pointwise* for all z in a certain subdisk. A similar statement can be made for derivatives. For these results it is essential that f be univalent.

Theorem. *If* $f \in S$, $g \prec f$, *and* $g'(0) \geq 0$, *then* $|g(z)| \leq |f(z)|$ *throughout the disk* $|z| \leq \frac{1}{2}(3-\sqrt{5}) = 0.381\ldots$. *This radius is best possible.*

This theorem has a long history. M. Biernacki [3] first proved it for $|z| \leq \frac{1}{4}$. Goluzin [14] improved the radius to 0.35 and conjectured the sharp radius to be $\frac{1}{2}(3-\sqrt{5})$. This he proved under the additional assumption that f is starlike. Goluzin's method is quite natural, yet apparently incapable of yielding the sharp result. Shah Tao-shing [28] introduced a more technical approach to establish Goluzin's conjecture. We shall give a brief sketch of Shah's argument.

Rogosinski's lemma shows it is sufficient (in fact, equivalent) to prove that $|f(\zeta)| \leq |f(z)|$ for all $\zeta \in \Delta_z$. If $|\zeta| \leq |z|^2 = r^2$, then by the classical estimates of functions in S,

$$|f(\zeta)| \leq \frac{r^2}{(1-r^2)^2} \qquad \text{and} \qquad \frac{r}{(1+r)^2} \leq |f(z)|\,.$$

Thus $|f(\zeta)| \leq |f(z)|$ if $r^2(1+r)^2 \leq r(1-r^2)^2$, or if $r^2 - 3r + 1 \geq 0$. This inequality holds precisely for $r \leq \frac{1}{2}(3-\sqrt{5})$.

The same calculation shows the radius cannot be increased beyond $\frac{1}{2}(3-\sqrt{5})$. Simply take f to be the Koebe function $k(z) = z(1-z)^{-2}$ and let $g(z) = k(z^2)$.

The real difficulty is to extend the estimate $|f(\zeta)| \leq |f(z)|$ to the part of Δ_z outside the disk $|\zeta| \leq |z|^2$. If $|f(\zeta)| > |f(z)|$ somewhere in this region, then $|f(\zeta)| > |f(z)|$ somewhere on one of the boundary arcs joining $\pm i|z|z$ to z. It would then follow by continuity that $|f(\zeta)| = |f(z)|$ for some point interior to this arc. Shah obtains a contradiction by appeal to an inequality of Lebedev for odd functions $h \in S$ (see Goluzin [16, p. 122]):

$$\left|\log\left(\frac{h(u)+h(v)}{h(u)-h(v)} \cdot \frac{u-v}{u+v}\right)\right|^2 \leq \log\frac{1+|u|^2}{1-|u|^2}\,\log\frac{1+|v|^2}{1-|v|^2}\,,$$

choosing h to be the square-root transform of f and letting $u = \sqrt{z}$ and $v = \sqrt{\zeta}$. The proof is highly technical and not very illuminating. It would be desirable to find a more conceptual proof.

There is a corresponding theorem for derivatives.

Theorem. *If* $f \in S$, $g \prec f$, *and* $g'(0) \geq 0$, *then* $|g'(z)| \leq |f'(z)|$ *throughout the disk* $|z| \leq 3-\sqrt{8} = 0.171\ldots$ *This radius is best possible.*

Shah [29] obtained this result by a similar but even more technical argument, making use of Dieudonné's lemma. (Again it would be desirable to find a more elegant proof.) Goluzin [14] had proved it for $|z| \leq 0.12$ and had conjectured the sharp bound. Goluzin also showed the radius $3-\sqrt{8}$ to be best possible by means of the example $f(z) = z(1+z)^{-2}$ and $g(z) = f(\omega(z))$, where

$$\omega(z) = z\,\frac{z+\alpha}{1+\alpha z}$$

and $\alpha < 1$ is sufficiently near to 1.

Z. Lewandowski [17] has established a partial converse, to the effect that pointwise domination ("majorization") implies subordination. Specifically, he proved that if $f \in S$ and if g is an analytic function with $g'(0) \geqslant 0$ for which $|g(z)| \leqslant |f(z)|$ in $|z| < 1$, then g is subordinate to f in the disk $|z| \leqslant 0.21$; *i.e.*, $g(D_r) \subset f(D_r)$ for all $r \leqslant 0.21$, where D_r is the disk $|z| < r$. The best possible radius is unknown, but Lewandowski gave a simple example to show it is no larger than 0.3. Lewandowski [18] and T. H. MacGregor [20] also showed that under the same hypotheses, $|g'(z)| \leqslant |f'(z)|$ in $|z| \leqslant 2 - \sqrt{3}$, the radius being best possible.

D. M. Campbell [4, 5, 6] has generalized most of the preceding results to the case in which f is locally univalent and belongs to a linearly invariant family of prescribed order.

Univalent subordinate functions.

The theorem on pointwise domination of $|g(z)|$ by $|f(z)|$ can be slightly improved under the additional hypothesis that g is univalent. Biernacki [2, 3] first obtained this result by a variational method, and Goluzin [12, 15] later based a proof on a variant of Loewner's method. Goluzin's approach is particularly well suited to problems of this type, in which the subordinate function is univalent. The result is as follows.

Theorem. *Suppose* $f \in S$, *and let* g *be analytic and univalent in the unit disk, with* $g \prec f$ *and* $g'(0) > 0$. *Then* $|g(z)| \leqslant |f(z)|$ *throughout the disk* $|z| \leqslant r_0 = 0.390\ldots$, *where* r_0 *is the unique solution to the transcendental equation*

$$\log\frac{1+r}{1-r} + 2\tan^{-1} r = \frac{\pi}{2} .$$

The radius r_0 *is best possible.*

This theorem suggests that the pointwise domination $|g'(z)| \leqslant |f'(z)|$ of derivatives should be similarly extendible beyond the radius $3 - \sqrt{8}$ under the additional hypothesis that g is univalent. Goluzin [15] also applied the Loewner method to this problem. The result is astonishing: $|g'(z)| \leqslant |f'(z)|$ for $|z| \leqslant 3 - \sqrt{8}$, and the radius is best possible. In other words, the univalence assumption on g is quite irrelevant and does not strengthen the conclusion.

This last result can also be deduced from Shah's more general theorem by producing an example to show the radius cannot be increased even if g is univalent. Goluzin [15] offers the following construction. Let $f(z) = z(1+z)^{-2}$ and let $g(z) = f(\omega(z))$, where

$$\omega(z) = k^{-1}((1-\epsilon)k(z)) , \qquad 0 < \epsilon < 1 ,$$

and $k(z) = z(1-z)^{-2}$. Then g is univalent, and a calculation gives

$$|g'(z)| = |f'(z)| \left[1 - \epsilon \operatorname{Re}\left\{\frac{z^2 - 6z + 1}{(1+z)^2}\right\} + O(\epsilon^2)\right] .$$

It follows that if $3 - \sqrt{8} < r < 1$, then $|g'(r)| > |f'(r)|$ for sufficiently small $\epsilon > 0$, since $r^2 - 6r + 1 < 0$ in this case.

References

1. A. Baernstein, Integral means, univalent functions and circular symmetrization, *Acta Math.* **133** (1974), 139–169.

2. M. Biernacki, Sur quelques majorantes de la théorie des fonctions univalentes, *C. R. Acad. Sci. Paris* **201** (1935), 256–258.

3. M. Biernacki, Sur les fonctions univalentes, *Mathematica (Cluj)* **12** (1936), 49–64.

4. D. M. Campbell, Majorization-subordination theorems for locally univalent functions, *Bull. Amer. Math. Soc.* **78** (1972), 535–538.

5. D. M. Campbell, Majorization-subordination theorems for locally univalent functions, II, *Canad. J. Math.* **25** (1973), 420–425.

6. D. M. Campbell, Majorization-subordination theorems for locally univalent functions, III, *Trans. Amer. Math. Soc.* **198** (1974), 297–306.

7. C. Carathéodory, Theory of Functions of a Complex Variable, Vol. II, *Chelsea translation*, New York, 1954.

8. J. Dieudonné, Recherches sur quelques problèmes relatifs aux polynomes et aux fonctions bornées d'une variable complexe, *Ann. Ecole Norm. Sup.* **48** (1931), 247–358.

9. P. L. Duren, **Theory of H^p Spaces**, Academic Press, New York, 1970.

10. P. L. Duren, **Univalent Functions**, Springer-Verlag, New York-Heidelberg, to appear.

11. S. Friedland, On a conjecture of Robertson, *Arch. Rational Mech. Anal.* **37** (1970), 255–261.

12. G. M. Goluzin, On the theory of univalent functions, *Mat. Sb.* **6** (48) (1939), 383–388. (in Russian)

13. G. M. Goluzin, On subordinate univalent functions, *Trudy Mat. Inst. Steklov.* **38** (1951), 68–71. (in Russian)

14. G. M. Goluzin, On majorants of subordinate analytic functions, I, *Mat. Sb.* **29** (71) (1951), 209–224. (in Russian)

15. G. M. Goluzin, On majorants of subordinate analytic functions, II, *Mat. Sb.* **29** (71) (1951), 593–602. (in Russian)

16. G. M. Goluzin, **Geometric Theory of Functions of a Complex Variable**, Moscow, 1952; German transl., Deutscher Verlag, Berlin, 1957; 2nd ed., Moscow, 1966; English transl., Amer. Math. Soc., 1969.

17. Z. Lewandowski, Sur les majorantes des fonctions holomorphes dans le cercle $|z| < 1$, *Ann. Univ. Mariae Curie-Skłodowska* **15** (1961), 5–11.

18. Z. Lewandowski, Some results concerning univalent majorants, *Ann. Univ. Mariae Curie-Skłodowska* **18** (1964), 13–18.

19. J. E. Littlewood, On inequalities in the theory of functions, *Proc. London Math. Soc.* **23** (1925), 481–519.

20. T. H. MacGregor, Majorization by univalent functions, *Duke Math. J.* **34** (1967), 95–102.

21. M. S. Robertson, A remark on the odd schlicht functions, *Bull. Amer. Math. Soc.* **42** (1936), 366–370.

22. M. S. Robertson, The generalized Bieberbach conjecture for subordinate functions, *Michigan Math. J.* **12** (1965), 421–429.

23. M. S. Robertson, Quasi-subordination and coefficient conjectures, *Bull. Amer. Math. Soc.* **76** (1970), 1–9.

24. M. S. Robertson, Quasi-subordinate functions, *Mathematical Essays Dedicated to A. J. MacIntyre* (Ohio Univ. Press, Athens, Ohio, 1970), pp. 311–330.

25. W. Rogosinski, Zum Schwarzschen Lemma, *Jber. Deutsch. Math.-Verein.* **44** (1934), 258–261.

26. W. Rogosinski, On subordinate functions, *Proc. Cambridge Philos. Soc.* **35** (1939), 1–26.

27. W. Rogosinski, On the coefficients of subordinate functions, *Proc. London Math. Soc.* **48** (1943), 48–82.

28. Shah Tao-shing, Goluzin's number $(3-\sqrt{5})/2$ is the radius of superiority in subordination, *Science Record* **1** (1957), no. 4, 25–28.

29. Shah Tao-shing, On the radius of superiority in subordination, *Science Record* **1** (1957), no. 5, 53–57.

Department of Mathematics
University of Michigan
Ann Arbor, Michigan 48109

EQUIDISTRIBUTION OF THE ARGUMENTS OF THE ZEROS OF PADÉ NUMERATORS

Albert Edrei

Introduction. Let

(1) $$f(z) = a_0 + a_1 z + a_2 z^2 + \cdots \qquad (a_0 \neq 0)\,,$$

be a power series having a positive radius of convergence R_0.

Let (m,n) be a pair of nonnegative integers. We say that the polynomials $P_{mn}(z), Q_{mn}(z)$ are *normalized Padé polynomials* of the entry (m,n) of the Padé table of (1) if the three following conditions are satisfied.

I. $Q_{mn}(z)$ *is a polynomial of degree* $\leq n$, *and*

(2) $$Q_{mn}(z) = z^k T_{mn}(z) \qquad (k = k(m,n) \geq 0,\ T_{mn}(0) = 1)$$

where T *is a polynomial.*

II. $P_{mn}(z)$ *is a polynomial of degree* $\leq m$.

III. $f(z)Q_{mn}(z) - P_{mn}(z) = z^{m+n+1}F_{mn}(z)$, *where* $F_{mn}(z)$ *is regular for* $|z| < R_0$.

It is well-known [3, pp. 235–237] that, for every entry (m,n), the normalized Padé polynomials exist and their ratio

(3) $$\frac{P_{mn}(z)}{Q_{mn}(z)}$$

is uniquely determined.

Our study is restricted to fixed values of the integer $n \geq 0$. This enables us to omit the suffix n from our notation for P and Q.

Analytic character of $f(z)$. *Throughout this note* $f(z)$ *is meromorphic in the region*

$$|z| < R_1 \qquad (0 < R_1 \leq +\infty)\,.$$

If $R_1 = +\infty$, *the function is meromorphic; if* $R_1 < +\infty$, *there is a non-polar singularity on the circumference* $|z| = R_1$.

We consider the sequence

(4) $$b_1, b_2, b_3, \ldots$$

The research of the author was supported by a grant from the National Science Foundation MCS 72-04539 A04.

of *all the poles* of $f(z)$ in $|z| < R_1$. Multiple poles are repeated as often as indicated by their multiplicities, and the sequence arranged so that

$$0 < |b_1| \leqslant |b_2| \leqslant |b_3| \leqslant \cdots .$$

My first result is

Theorem 1. [Extension of the theorem of de Montessus de Ballore.] *Let* $f(z)$ *be given by* (1); *let* R $(0 < R < +\infty)$ *be such that the only singularities of* $f(z)$, *in* $|z| < R$, *are* $q \geqslant 0$ *poles*

(5) $$b_1, b_2, \ldots, b_q \qquad (|b_q| < R) ,$$

and let n *be a given integer*

(6) $$n \geqslant q .$$

Introduce the disks

$$\Gamma_j(\rho): |z - b_j| \leqslant \rho \qquad (0 < \rho, j = 1,2, \ldots, q) ,$$

where ρ *is so small that*

$$\{\Gamma_j(\rho) \cap \Gamma_k(\rho)\} = 0 \qquad (b_j \neq b_k) ,$$

otherwise arbitrary.

Then, if $m > m_0(n,\rho)$, *each disk* $\Gamma_j(\rho)$ *contains at least as many zeros of* $Q_m(z)$ *as indicated by the multiplicity of* b_j.

The above theorem readily yields

Corollary 1. *From the sequence of positive integers it is possible to extract an infinite subsequence* $\mathcal{M}$ *such that, in* $|z| < R$,

(7) $$\frac{P_m(z)}{Q_m(z)} \to f(z) \qquad (m \to \infty, m \in \mathcal{M}) ,$$

except at $p \leqslant n$ *exceptional points*

(8) $$e_1, e_2, \ldots, e_p .$$

The set (8) *contains the* q *poles of* (5) *and at most* $n - q$ *other points.*

The convergence in (7) *is uniform on every compact subset of* $|z| < R$ *which omits the points of* (8).

In the special case $q = n$, the polynomial $Q_m(z)$ which (by definition) has at most n zeros, must, by Theorem 1, have at least n zeros, properly distributed in the disks $\Gamma_j(\rho)$.

Putting

(9) $$V_\ell(z) = \prod_{j=1}^{\ell} \left(1 - \frac{z}{b_j}\right) \qquad (\ell = 0,1,2, \ldots) ,$$

and letting $\rho \to 0$, we thus obtain

$$(10) \qquad Q_m(z) \to V_q(z), \; P_m(z) \to f(z)V_q(z) \qquad (m \to \infty),$$

uniformly on every compact subset of the disk $|z| < R$.

The latter statement is one form of the classical theorem of de Montessus de Ballore [3, p. 265], [1, pp. 139–142].

The proof of Theorem 1 may be of interest for several reasons:

(i) it does not involve determinants and uses instead the rudiments of function theory;

(ii) it leads to a simple, "painless" proof of the theorem of de Montessus de Ballore and suggests the possibility of treating, in the same way, some interesting extensions due to R. Wilson [7], [8], [9], [10], [1, pp. 143–149];

(iii) it is a useful tool in the proof of my main result, stated below as Theorem 2.

It should also be remarked that Theorem 1 provides some information in cases not covered by the theorem of de Montessus de Ballore, for instance, in the case of entire functions or of functions without poles in the disk $|z| < R_1$ $(R_1 < +\infty)$.

Theorem 2. *Let* $f(z)$ *have the analytic character described above and let the fixed integer* $n \geqslant 0$ *be such that* b_{n+1} *exists and*

$$(11) \qquad |b_n| < |b_{n+1}| .$$

It is then possible to find an infinite sequence M *of positive, strictly increasing integers such that the sequence of Padé numerators*

$$P_m(z) \qquad (m \in M)$$

behaves as follows.

I. $P_m(0) = a_0$.

II. *The degree of* $P_m(z)$ *is exactly* m.

III. *Put*

$$P_m(z) = a_0 \prod_{j=1}^{m} \left(1 - \frac{z}{\omega_j}\right) ,$$

and

$$\omega_j = \omega_j(m) = |\omega_j| e^{i\theta_j} \qquad (\theta_j = \theta_j(m)) .$$

Then the number of zeros of $P_m(z)$, *in every annulus*

$$|b_{n+1}| - \epsilon \leqslant |z| \leqslant |b_{n+1}| + \epsilon \qquad (0 < \epsilon < |b_{n+1}|, \; \epsilon \text{ fixed}),$$

is equal to $(1 - \eta_m)m$, *where* $\eta_m \to 0$ *as* $m \to \infty$, $m \in M$.

IV. *The arguments of the zeros of* $P_m(z)$ *are equidistributed in the sense of Weyl, that is, if* $N(m;\alpha,\beta)$ *denotes the number of* $\theta_j(m)$ *in the angle*

$$\alpha \leqslant \arg z \leqslant \beta \qquad (\alpha < \beta < \alpha + 2\pi),$$

we have

$$\frac{N(m;\alpha,\beta)}{m} \rightarrow \frac{\beta-\alpha}{2\pi} \qquad (m\rightarrow\infty,\ m\in M).$$

In addition to the methods developed in this note, the proof of Theorem 2 depends on some results of Rosenbloom [4].

A proof of Theorem 2 would inordinately lengthen this note. I shall omit it and expect to publish it elsewhere. I do not believe that it is necessary to impose on n a condition such as (11) and hope that a closer analysis will enable me to remove this restriction.

1. The main lemmas. Here and throughout the remainder of this note, K denotes a finite positive quantity which may depend on several parameters but *never* on the integer m or on the argument θ. The value of K need not be the same one at each occurence; whenever clarity demands it, we shall identify the various values of K by writing $K_1, K_2, \ldots$.

The normalized Padé denominator $Q_{mn}(z)$ is of the form

$$Q_{mn}(z) = z^k \prod_j \left(1-\frac{z}{\zeta_j}\right) \qquad (\zeta_j = \zeta_j(m,n);\ k = k(m,n) \geqslant 0), \tag{1.1}$$

and it is convenient to arrange its zeros so that

$$0 < |\zeta_1| \leqslant |\zeta_2| \leqslant \cdots .$$

Similarly, the normalized Padé numerator $P_{mn}(z)$ is

$$P_{mn}(z) = a_0 z^k \prod_j \left(1-\frac{z}{\omega_j}\right) \qquad (0<|\omega_1|\leqslant|\omega_2|\leqslant\cdots). \tag{1.2}$$

The notations of (1.1) and (1.2) are adopted throughout this note.

Lemma 1. *Let $f(z)$ be given by (1) and let $f(z)$ have the analytic character described in the introduction.*

Assume that $f(z)$ is regular in the disk $|z|<t'$ except for s poles

$$b_1, b_2, \ldots, b_s \qquad (s\geqslant 0,\ 0<|b_1|\leqslant|b_s|<t'). \tag{1.3}$$

Select any t such that

$$|b_s| < t < t',$$

and let $\eta>0$ be such that

$$|b_s|+\eta < t < t+\eta < t', \tag{1.4}$$

otherwise arbitrary.

The positive integer n is fixed,

$$n \geqslant s, \tag{1.5}$$

and $Q_m(z), P_m(z)$ are the abbreviated notations for $Q_{mn}(z), P_{mn}(z)$.

Assume also that for some particular value of $m > m_0 = m_0(t,\eta,n)$ *it is possible to find a complex quantity* z_0 *such that*

(1.6) $$|z_0| < t - \eta$$

and such that the disk

$$\Delta^* = \{z\colon |z - z_0| \leq 2\eta\},$$

contains no zero of $Q_m(z)$ *and none of the poles* (1.3).

Then

(1.7) $$\left| f(z) - \frac{P_m(z)}{Q_m(z)} \right| \leq K_1 \left(\frac{t}{t+\eta}\right)^m ,$$

throughout the disk

$$\Delta = \{z\colon |z - z_0| \leq \eta\} .$$

The constant $K_1 = K_1(t,\eta,n)$ *is independent of* m *and* z.

Proof. Let $V(z) = V_s(z)$ be the polynomial defined in (9) and notice that Cauchy's Theorem implies [2, p. 436]

(1.8) $$V(z)\{f(z)Q_m(z) - P_m(z)\} = \frac{z^{m+n+1}}{2\pi i} \int_C \frac{f(w)Q_m(w)V(w)}{w^{m+n+1}(w-z)}\, dw \qquad (|z| < t + \eta) ,$$

where the integration is over the circumference

$$C\colon\ w = (t+\eta)e^{i\theta} \qquad (0 \leq \theta < 2\pi) .$$

[The validity of the relation (1.8) requires the assumption (1.5).]

Since

$$|z_0 - \zeta_j| > 2\eta , \qquad |z_0 - b_j| > 2\eta ,$$

the relation $z \in \Delta$ implies

(1.9) $$|\zeta_j - z| > \eta \quad (j = 1,2,3,\ldots) ; \qquad |b_j - z| > \eta \quad (j = 1,2,\ldots,s) .$$

Also, by (1.6),

(1.10) $$|z| < t .$$

Hence, if

(1.11) $$|w| = t + \eta , \qquad z \in \Delta ,$$

we find

(1.12) $$\left|\frac{Q_m(w)}{Q_m(z)}\right| = \left|\frac{w}{z}\right|^k \prod_{0<|\zeta_j|\leq 2|w|} \left|\frac{w-\zeta_j}{z-\zeta_j}\right| \prod_{|\zeta_j|>2|w|} \left|\frac{1-w/\zeta_j}{1-z/\zeta_j}\right| \leq \left|\frac{w}{z}\right|^k 3^n \left|\frac{w}{\eta}\right|^n = K_2 \left|\frac{w}{z}\right|^k .$$

Now

$$\max_{|z|\leqslant|w|} |f(z)V(z)| = K_3 , \qquad \min_{z\in\Delta} |V(z)| = K_4 ;$$

using these relations and (1.12) in (1.8) we obtain

$$\left| f(z) - \frac{P_m(z)}{Q_m(z)} \right| \leqslant \frac{K_2 K_3}{K_4} \left|\frac{w}{\eta}\right| \left(\frac{t}{t+\eta}\right)^{m+n+1-k} \qquad (z \in \Delta) ,$$

and (1.7) follows.

Lemma 2. *Let* $f(z)$ *have the poles* b_j *listed in* (1.3), *and let* t', t, η *and* n *satisfy the conditions* (1.4) *and* (1.5).

Assume, in addition, that

$$f(te^{i\theta}) \neq 0 \qquad (0 \leqslant \theta \leqslant 2\pi) , \tag{1.13}$$

that $f(z)$ *has exactly* ℓ *zeros in* $|z| < t$ *and that* η *is small enough to imply*

$$K \leqslant |f(re^{i\theta})| \leqslant K \qquad (0 \leqslant \theta < 2\pi) , \tag{1.14}$$

throughout the annulus

$$A : t - \eta \leqslant r \leqslant t . \tag{1.15}$$

Then, with every $m > m_0$, *it is possible to associate a radius* ρ_m *displaying the following behavior.*

A. $t - \eta < \rho_m < t$.

B. $P_m(z)$ *and* $Q_m(z)$ *have no zeros on the circumference* $|z| = \rho_m$.

C. *Let* $\lambda(m)$ *and* $\sigma(m)$ *denote, respectively, the number of zeros of* $P_m(z)$ *and* $Q_m(z)$, *in* $|z| < \rho_m$. *Then*

$$\ell - s = \lambda(m) - \sigma(m) . \tag{1.16}$$

D. *For suitable values of the constants* K:

$$\left| f(\rho_m e^{i\theta}) - \frac{P_m(\rho_m e^{i\theta})}{Q_m(\rho_m e^{i\theta})} \right| \leqslant K \left(\frac{t}{t+\eta}\right)^m \qquad (0 \leqslant \theta < 2\pi) , \tag{1.17}$$

and

$$K \leqslant \frac{\prod_{|\zeta_j| \neq 0,\, j \leqslant \sigma} |\zeta_j|}{\prod_{|\omega_j| \neq 0,\, j \leqslant \lambda} |\omega_j|} \leqslant K \qquad (\sigma = \sigma(m),\ \lambda = \lambda(m)\ ) . \tag{1.18}$$

Proof. Consider disks such as

$$|z - \zeta_j| \leqslant \frac{\eta}{8(n+1)} , \qquad |z - b_j| \leqslant \frac{\eta}{8(n+1)} ,$$

there are no more than $2n$ of them and the sum of all the diameters of these "exceptional" disks does not exceed $n\eta/2(n+1)$. Hence it is possible to select ρ_m such that

(1.19) $$t - \eta < \rho_m < t - \frac{\eta}{2}$$

and such that the circumference $|z| = \rho_m$ does not intersect any of the exceptional disks.

Let $|z_0| = \rho_m$; by construction, the distance between z_0 and any one of the ζ_j or b_j exceeds

$$\frac{\eta}{8(n+1)} = 2\xi$$

and, by (1.19),

$$\rho_m < t - \xi .$$

An inspection of (1.12) reveals that we now have

$$\left| \frac{Q_m(w)}{Q_m(z_0)} \right| \leqslant \left| \frac{w}{\rho_m} \right|^k 3^n \left| \frac{w}{\xi} \right|^n \qquad (|w| = t + \eta,\ |z_0| = \rho_m),$$

and the proof of Lemma 1 shows that ρ_m satisfies the relation (1.17). As to assertion (A), it follows from (1.19).

In view of (1.14) and (1.17), we verify assertion (B) of the lemma. If we also use Rouché's lemma for meromorphic functions [5, p. 193], we obtain assertion (C).

To prove (1.18), denote by $\tau(m)$ the ratio of ζ's and ω's which constitutes the central term of this double inequality.

Notice that

$$\lim_{z \to 0} \frac{P_m(z)}{Q_m(z)} = a_0 ,$$

and consequently, Jensen's theorem yields

(1.20) $$\log |a_0| + \log \left(\rho_m^{\lambda-\sigma} \tau(m) \right) = \frac{1}{2\pi} \int_0^{2\pi} \log \left| \frac{P_m(\rho_m e^{i\theta})}{Q_m(\rho_m e^{i\theta})} \right| d\theta .$$

In view of (1.14), (1.16), (1.17) and (1.19), we immediately see that (1.20) implies (1.18). The proof of Lemma 2 is now complete.

2. Three normal families. We now assume that all the conditions of Theorem 1 are satisfied, and consider the factored forms (1.1) and (1.2) of the Padé polynomials.

Take h $(0 < 2h < R)$ so small that $f(z)$ has no zeros and no poles in $|z| \leqslant 2h$.

A first application of Lemma 2 enables us to determine a sequence $\{h_m\}$ such that

(2.1) $$\frac{h}{2} < h_m < h \qquad (m > m_0)$$

and such that there are in $|z| < h_m$ exactly as many zeros of $Q_m(z)$ as there are zeros of $P_m(z)$, say $k + \nu$ $(k = k(m),\ \nu = \nu(m))$ of them.

Consider the family of rational functions

$$\Lambda_m(z) = \prod_{j=1}^{\nu} \frac{1 - z/\omega_j}{1 - z/\zeta_j} \qquad (\nu = \nu(m)) . \tag{2.2}$$

For

$$|z| \geqslant 2h , \qquad m > m_o , \tag{2.3}$$

we must have, by (1.18),

$$K\,3^{-n} \leqslant |\Lambda_m(z)| \leqslant K\,3^n . \tag{2.4}$$

Consider next two families of polynomials

$$\psi_m(z) = \prod_{j>\nu} \left(1 - \frac{z}{\zeta_j}\right) , \qquad \varphi_m(z) = \prod_{j>\nu} \left(1 - \frac{z}{\omega_j}\right) , \tag{2.5}$$

and notice the obvious inequality

$$|\psi_m(z)| \leqslant \left(1 + \frac{2|z|}{h}\right)^n . \tag{2.6}$$

To study $\varphi_m(z)$ we apply Lemma 2 once more, with s replaced by q. We select H, H' such that

$$\max\,\{2h, |b_q|\} < H < H' < R$$

and then a sequence $\{H_m\}$ such that

$$H < H_m < H' \qquad (m > m_o) \tag{2.7}$$

and such that (1.17) holds in the form

$$\left| f(H_m e^{i\theta}) - \frac{P_m(H_m e^{i\theta})}{Q_m(H_m e^{i\theta})} \right| \leqslant K\chi^m \qquad (0 \leqslant \theta < 2\pi,\ m > m_o) , \tag{2.8}$$

where χ $(0 < \chi < 1)$ is a suitable constant (depending on H and H'), independent of m and θ.

From (2.8), (2.2) and (2.5), we deduce

$$\left| \Lambda_m(H_m e^{i\theta}) \frac{\varphi_m(H_m e^{i\theta})}{\psi_m(H_m e^{i\theta})} \right| \leqslant |f(H_m e^{i\theta})| + K ,$$

and, in view of (2.3), (2.4) and (2.6),

$$|\varphi_m(H_m e^{i\theta})| \leqslant K\,3^n \left(1 + \frac{2R}{h}\right)^n \left\{ \max_{0 \leqslant \theta < 2\pi} |f(H_m e^{i\theta})| + K \right\} . \tag{2.9}$$

Since $\varphi_m(z)$ is a polynomial, (2.7), (2.9), and the maximum modulus principle yield

$$\max_{|z| \leqslant H} |\varphi_m(z)| \leqslant K \qquad (m > m_o) . \tag{2.10}$$

3. Proof of Theorem 1. In this section M denotes an infinite subsequence of the sequence of positive integers and M_1 an infinite subsequence of M.

Assume that Theorem 1 is false. It is then possible to find

$$b_\gamma, \rho \quad (0 < \rho < |b_\gamma|, \ |b_\gamma| + \rho < R), \ M,$$

such that if $m \in M$, the polynomial $Q_m(z)$ has exactly μ zeros in $\Gamma_\gamma(\rho)$ and

(3.1) $$\mu < \mu_\gamma = \text{multiplicity of pole } b_\gamma .$$

Using the results of section 2, we select h and H such that

$$0 < h < |b_\gamma| - \rho, \qquad |b_\gamma| + \rho < H < R,$$

and such that the three families

(3.2) $$\{\psi_m(z)\}, \quad \{\varphi_m(z)\}, \quad \{\Lambda_m(z)\} \qquad (m \in M),$$

are uniformly bounded in the annulus

$$\widetilde{A}: h \leqslant |z| \leqslant H .$$

The families (3.2) are normal in $\widetilde{A}$ and consequently we may select M_1 such that as $m \to \infty$, $m \in M_1$, we have

(3.3) $$\psi_m(z) \to \psi(z), \qquad \varphi_m(z) \to \varphi(z), \qquad \Lambda_m(z) \to \Lambda(z),$$

uniformly on every compact subset $S \subset \widetilde{A}$.

Since $\psi_m(z)$ is uniformly bounded in $|z| \leqslant H$, we know, in addition, that the restriction $|z| \geqslant h$ may be omitted in the relation

$$\psi_m(z) \to \psi(z) .$$

This shows that $\psi(z) \not\equiv 0$ because $\psi_m(0) = 1$ implies $\psi(0) = 1$.

We now see that (3.3) implies

(3.4) $$f(z) - \frac{\Lambda(z)\varphi(z)}{\psi(z)} = g(z),$$

where $g(z)$ is regular for $z \in S$, except possibly at the zeros of $\psi(z)$ and at the poles of $f(z)$.

Consider now some closed disk $\widetilde{\Delta}$, which belongs to the interior of $\widetilde{A}$, and contains neither poles of $f(z)$ nor zeros of $\psi(z)$. In view of Hurwitz' theorem [6, p. 119] a disk Δ' having the same center as $\widetilde{\Delta}$, and a smaller radius, will contain no zeros of $\psi_m(z)$ for $m > m_0$, $m \in M_1$. We now inspect Lemma 1 and, in particular, compare (1.7) and (3.4). This reveals that in the neighborhood of the center of Δ', $g(z) \equiv 0$. By analytic continuation

(3.5) $$f(z) = \Lambda(z)\frac{\varphi(z)}{\psi(z)},$$

throughout the annulus $\widetilde{A}$. By (3.1) and Hurwitz' theorem, the right-hand side of (3.5) has, at b_γ, a pole of order $\leqslant \mu$, whereas, by assumption, b_γ is a pole of order $\mu_\gamma > \mu$ of $f(z)$. This contradiction forces us to reject the possibility of the behavior which leads to (3.1) and, consequently, proves Theorem 1.

References

1. G. A. Baker, Jr., **Essentials of Padé Approximants**, Academic Press, New York, 1975.
2. A. Edrei, The Padé table of functions having a finite number of essential singularities, *Pacific J. Math.* **56** (1975), 429–453.
3. O. Perron, **Die Lehre von den Kettenbrüchen**, 3rd. ed., vol. 2, B. G. Teubner, Stuttgart, 1957.
4. P. C. Rosenbloom, Sequences of polynomials, especially sections of power series, Thesis, Stanford, 1943.
5. S. Saks and A. Zygmund, **Analytic Functions**, Monografie Matematyczne, Warsaw, 1952.
6. E. C. Titchmarsh, **The theory of Functions**, 2nd. ed., Oxford University Press, Oxford, 1952.
7. R. Wilson, Divergent continued fractions and polar singularities, *Proc. London Math. Soc.* **26** (1927), 159–168.
8. R. Wilson, Divergent continued fractions and polar singularities, Part II, *Proc. London Math. Soc.* **27** (1928), 497–512.
9. R. Wilson, Divergent continued fractions and polar singularities, Part III, *Proc. London Math. Soc.* **28** (1928), 128–144.
10. R. Wilson, Divergent continued fractions and non-polar singularities, *Proc. London Math. Soc.* **30** (1930), 38–57.

Syracuse University
Syracuse, New York 13210

AN EXTREMAL PROBLEM IN FUNCTION THEORY

Matts Essén and Daniel F. Shea*

Let $f(z)$ be meromorphic and of order $\rho < 1$ in the plane, with deficiency $\delta(\infty,f) > 0$. We wish to compare the growth of the Nevanlinna characteristic $T(r,f)$ with that of

$$A(r) = A(r,f) = \inf_{|z|=r} \log |f(z)| .$$

For entire $f(z)$, Edrei and Fuchs found the inequality

$$\limsup_{r\to\infty} \frac{A(r,f)}{T(r,f)} \geqslant \pi\rho \cos \pi\rho \qquad (\tfrac{1}{2} \leqslant \rho < 1) , \tag{1}$$

see [5], which is sharp for the Lindelöf functions:

$$f_\rho(z) = \prod_{n=1}^{\infty} (1 + z/n^{1/\rho}) .$$

In conversation, Edrei asked how it could be shown that the $f(z)$ "extremal" for (1), in the sense that (1) holds with equality, behave asymptotically like Lindelöf functions. A result of this type was established in [5] for the analogue of (1) for orders $\rho < \tfrac{1}{2}$, see also [3], [4], [10] for other, similar, results. Our purpose here is to indicate a method that yields the desired information, based on the convolution inequality

$$T(r) \geqslant -A * K(r) + S_\beta(r) \tag{2}$$

where

$$A * K(r) = \int_0^\infty A(t) K\left(\frac{r}{t}\right) \frac{dt}{t} ,$$

$$K(r) = \frac{1}{\pi^2} \log \left| \frac{1-\zeta}{1+\zeta} \right| + \frac{2}{\pi^2} \operatorname{Re}(\zeta) \qquad (\zeta = r^{1/2} e^{i(\pi-\beta)/2}) , \tag{3}$$

$\beta = \pi/2\rho$ and S_β is an error term to be described below.

One might hope to base a study of the functions extremal for (1) on the known convolution inequality

$$A(r) \geqslant -T * k(r) - \Sigma(r) , \qquad k(r) = r^{2\rho}/(1 + r^{2\rho})^2 , \tag{4}$$

where Σ is defined in terms of the zeros of $f(z)$ in a certain angle $A(r)$ (see Fuchs [7], Petrenko [9]), but this seems unlikely since (4) is sharp only for meromorphic $f(z)$ having few zeros in $A(r)$, $0 < r < \infty$. Our inequality (2) takes into account the behavior of $f(z)$ in the full plane.

The method to be described applies as well to meromorphic functions f of order $\rho \in [\tfrac{1}{2},1)$ and deficiency $\delta(\infty,f) > 1 - \sin \pi\rho$, which are known ([6], [8]) to satisfy

---

* Research supported in part by a grant from the Swedish Natural Science Research Council.

(5) $$\limsup_{r\to\infty} \frac{A(r)}{T(r)} \geqslant \Gamma(\rho, \delta(\infty,f)) ,$$

$$\Gamma(\rho,\delta) = \pi\rho\{\sqrt{\delta(2-\delta)}\cos\pi\rho - (1-\delta)\sin\pi\rho\} .$$

This result is contained in

Theorem 1. *Let* $f(z)$ *be extremal for* (5), *i.e., suppose* f *has order* ρ *and*

(6) $$A(r) \leqslant \{\Gamma(\rho,\delta(\infty)) + o(1)\}T(r) \qquad (r\to\infty)$$

where $\tfrac{1}{2} \leqslant \rho < 1$ *and* $\delta(\infty) = \delta(\infty,f) > 1 - \sin\pi\rho$.

Then there exists $G = \bigcup_{n=1}^{\infty} [a_n,b_n]$ *where* $a_n \to \infty$, $b_n/a_n \to \infty$ *and*

(7) $$\lim_{r\to\infty} \frac{1}{\log r} \int_{G\cap[1,r]} d(\log t) = 1 ,$$

such that $L(r)$ *defined by* $T(r) = r^{\rho}L(r)$ *satisfies*

(8) $$\lim_{\substack{r\to\infty \\ r\in G}} \frac{L(\sigma r)}{L(r)} = 1 \qquad (0<\sigma<\infty) ,$$

and

(9) $$\lim_{\substack{r\to\infty \\ r\in G}} \frac{N(r,0)}{T(r)} = u , \qquad \lim_{\substack{r\to\infty \\ r\in G}} \frac{N(r,\infty)}{T(r)} = v$$

where $v = 1-\delta(\infty)$ *and* $u \in (0, \sin\pi\rho]$ *is determined by*

(10) $$u^2 + v^2 - 2uv\cos\pi\rho = \sin^2\pi\rho ;$$

(6) *holds with equality when* $r\to\infty$ *in* $G - E$, *where* E *has linear density zero.*

Further, there exists a real-valued $\theta(r)$ *such that the zeros* $\{z_n\}$ *and poles* $\{w_n\}$ *of* $f(z)$ *satisfy*

(11) $$\sum_{\substack{|z_n|\leqslant kr \\ \eta\leqslant|\arg z_n-\theta(r)|\leqslant\pi}} 1 + \sum_{\substack{|w_n|\leqslant kr \\ |\arg w_n-\theta(r)|\leqslant\pi-\eta}} 1 = o(T(r)) \qquad (r\to\infty \text{ in } G)$$

for any constants $k>1$ *and* $\eta>0$, *and*

(12) $$\lim_{\substack{r\to\infty \\ r\in G}} [\theta(\sigma r) - \theta(r)] = 0 \qquad (0<\sigma<\infty) ,$$

uniformly for $k^{-1} \leqslant \sigma \leqslant k$.

We give an outline of the proof, with details to appear later. We first require an extension of (2),

(13) $$T^*(re^{i\beta}) = -A * K(r) - N_\infty * K_1(r) + S_\beta(r) ,$$

where $N_\infty(r) = N(r,\infty;f)$, K is defined as in (3) with β determined by

$$\cos\beta\rho = 1 - \delta(\infty,f) \qquad (0<\beta\leqslant\frac{\pi}{2\rho}),$$

$$S_\beta(r) = S_\beta(r,T^*) = \int_0^\pi\int_0^\infty \Delta_1 T^*(te^{i\varphi})\, W_\beta(r/t,\varphi)\,\frac{dt}{t}\,d\varphi$$

where

$$\Delta_1 = t^2\Delta = t^2\frac{\partial^2}{\partial t^2} + t\frac{\partial}{\partial t} + \frac{\partial^2}{\partial\varphi^2}$$

and

$$-2\pi W_\beta(r,\varphi) = \log\left|\frac{e^{i\varphi/2}+r^{1/2}e^{i\beta/2}}{e^{i\varphi/2}+r^{1/2}e^{-i\beta/2}}\right| + \log\left|\frac{e^{i\varphi/2}-r^{1/2}e^{-i\beta/2}}{e^{i\varphi/2}-r^{1/2}e^{i\beta/2}}\right| - 4r^{1/2}\sin(\frac{\beta}{2})\sin(\frac{\varphi}{2}),$$

$$K_1(r) = \frac{\partial}{\partial\varphi}W_\beta(r,\varphi)\Big|_{\varphi=0}.$$

Here we have assumed $f(0) = 1$, and interpret

$$S_\beta(r,T^*) = \lim_{n\to\infty} S_\beta(r,U_n)$$

where

$$U_n(te^{i\varphi}) = \frac{1}{\pi}\int_0^{2\pi}\int_0^\infty e^{-\tau^2}\,T^*\left(te^{i\varphi}+\left(\frac{\tau}{n}\right)e^{i\psi}\right)\tau\,d\tau\,d\psi$$

and $T^*(\zeta) = T^*(\bar\zeta)$ for $\operatorname{Im}\zeta<0$. Then $U_n\in C^\infty$ and $U_n\downarrow T^*$. To prove (13), apply Green's formula to $U_n(\zeta)$ and

$$V_z(\zeta) = \log\left|\frac{1+\sqrt{z/\zeta}}{1+\sqrt{\bar z/\zeta}}\right| + \log\left|\frac{1-\sqrt{\bar z/\zeta}}{1-\sqrt{z/\zeta}}\right| - 2\operatorname{Re}\left(\frac{\sqrt{z}-\sqrt{\bar z}}{\sqrt{\zeta}}\right)$$

$$(z = re^{i\beta},\ \zeta = te^{i\varphi})$$

on the half-disc $|\zeta|\leqslant R$, $0\leqslant\varphi\leqslant\pi$ (compare [7, p. 19]), and let $R\to\infty$, then $n\to\infty$.

Here, our choice of V_z is motivated by these properties of Baernstein's function T^*:

$$T(|z|)\geqslant T^*(z), \qquad \frac{\partial}{\partial\varphi}T^*(\zeta)\Big|_{\varphi=\pi} = \frac{1}{\pi}A(t), \qquad T^*(t) = N_\infty(t) \qquad (0<t<\infty),$$

see [1]. Other information contained in T^* would be intrusive in our problem and must be surpressed; this is accomplished through the properties

$$V_z(t)\equiv 0, \qquad \frac{\partial}{\partial\varphi}V_z(te^{i\varphi})\Big|_{\varphi=\pi}\equiv 0 \qquad (0<t<\infty),$$

$V_z(\zeta) + \log|\zeta - z|$ is harmonic in $\operatorname{Im}\zeta>0$.

The important fact about T^*, that $\Delta T^*\geqslant 0$, (see [2]), can only be exploited if W_β has good sign-properties. It is sufficient that, for $C(r) = r^\alpha$ $(0<r<1)$, $C(r) = 0$ $(r\geqslant 1)$ and an appropriate $\alpha\in(\rho,1]$, we have

$$(14) \qquad X_\beta(x,\varphi) \equiv W_\beta(r,\varphi) * C(x) = x^\alpha \int_x^\infty W_\beta(r,\varphi) \frac{dr}{r^{1+\alpha}} > 0$$

for all $x > 0$, $0 < \varphi < \pi$. Here we require the evaluation

$$\int_0^\infty W_\beta(r,\varphi) \frac{dr}{r^{1+\alpha}} = \begin{cases} \dfrac{\sin \beta\alpha \cos (\pi - \varphi)\alpha}{-\alpha \cos \pi\alpha} & (\beta \leqslant \varphi \leqslant \pi) \\[2ex] \dfrac{\cos (\pi - \beta)\alpha \sin \varphi\alpha}{-\alpha \cos \pi\alpha} & (0 \leqslant \varphi \leqslant \beta), \end{cases}$$

valid for $\frac{1}{2} < \alpha < \frac{3}{2}$.

Thus we are led to

$$T * C(x) \geqslant -A * K * C(x) - N_\infty * K_1 * C(x) \qquad (0 < x < \infty),$$

and using our hypothesis (6) and the definition of $\delta(\infty)$, we obtain

$$(15) \qquad T * C(x) \geqslant \{1 - o(1)\} T * H(x) \qquad (x \to \infty)$$

where

$$H = -\Gamma(\rho,\delta(\infty))K * C - (1 - \delta(\infty))K_1 * C .$$

Although (15) is different from the convolution inequalities studied in [3], the methods used there can be adapted to yield a set G having the properties stated in Theorem 1, on which $L(r) = T(r)/r^\rho$ satisfies (8). Further exploitation of (13) yields

$$T^*(re^{i\beta},f) \sim T(r), \quad N_\infty(r) \sim [1 - \delta(\infty)]\, T(r) \qquad (r \to \infty \text{ in } G),$$

$$A(r) = \{\Gamma(\rho,\delta(\infty)) + o(1)\} T(r) \qquad (r \to \infty \text{ in } G - E)$$

for E of linear density zero, and also

$$(16) \qquad \int_0^\pi \int_0^\infty \Delta_1 T^*(te^{i\varphi},f) X_\beta(x/t,\varphi) \frac{dt}{t}\, d\varphi = o(T(x)) \qquad (x \to \infty \text{ in } G).$$

The method indicated above for deriving (13) yields also, for any f of genus zero with $f(0) = 1$,

$$(17) \qquad T^*(re^{i\beta},f) = N_0 * P_\beta(r) + N_\infty * P_{\pi-\beta}(r) - \widetilde{S}_\beta(r)$$

where $N_0(r) = N(r,0;f)$,

$$P_\beta(r) = \frac{1}{\pi} \frac{r \sin \beta}{r^2 + 1 + 2r \cos \beta},$$

$$(18) \qquad \widetilde{S}_\beta(r) = \frac{1}{2\pi} \int_0^\pi \int_0^\infty \Delta_1 T^*(te^{i\varphi},f) \widetilde{W}_\beta(r/t,\varphi) \frac{dt}{t}\, d\varphi ,$$

$$\widetilde{W}_\beta(r,\varphi) = \log \left| \frac{e^{i\varphi} - re^{-i\beta}}{e^{i\varphi} - re^{i\beta}} \right| .$$

It follows easily from (16) that

$$\widetilde{S}_\theta(r) = o(T(r)) \qquad (r \to \infty \text{ in } G,\ 0 < \theta < \pi),$$

and now known tauberian methods yield (9)–(12), cf. [3], [10].

If we introduce

(19) $$F(z) = \frac{\Pi(1 + z/|z_n|)}{\Pi(1 - z/|w_n|)},$$

where $\{z_n\}$ and $\{w_n\}$ are the zeros and poles of $f(z)$, then (17)–(18) give a useful expression for the difference

(20) $$\widetilde{S}_\theta(r) = T^*(re^{i\theta},F) - T^*(re^{i\theta},f) \qquad (0 < \theta < \pi).$$

The fact that $\widetilde{S}_\theta \geqslant 0$ is well known, and elementary.

It is easy to check directly that $T^*(re^{i\theta},F)$ is harmonic in the upper half-plane. Conversely, we deduce from (20):

Corollary. *Let* $f(z)$ *be meromorphic, of genus zero, with zeros* $\{z_n\}$, *poles* $\{w_n\}$ *and* $f(0) \neq 0,\infty$. *If* $T^*(re^{i\theta},f)$ *is harmonic, then*

$$f(z) = f(0)F(e^{i\alpha}z)$$

for some real α *and* $F(z)$ *as in* (19).

The analogous statement still holds when $f(z)$ has a zero or pole at the origin.

This result can be used with (16) and a compactness argument to give an alternate proof of (11)–(12).

References

1. A. Baernstein, A generalization of the cos $\pi\rho$ theorem, *Trans. Amer. Math. Soc.* **193** (1974), 181–197.

2. A. Baernstein and B. A. Taylor, Spherical rearrangements, subharmonic functions, and *-functions in n-space, *Duke Math. J.* **43** (1976), 245–268.

3. D. Drasin and D. F. Shea, Convolution inequalities, regular variation and exceptional sets, *J. Analyse Math.* **29** (1976), 232–293.

4. A. Edrei, Locally tauberian theorems for meromorphic functions of lower order less than one, *Trans. Amer. Math. Soc.* **140** (1969), 309–332.

5. A. Edrei, Extremal problems of the cos $\pi\rho$-type, *J. Analyse Math.* **29** (1976), 19–66.

6. M. Essén and D. F. Shea, Applications of Denjoy integral inequalities to growth problems for subharmonic and meromorphic functions, *Proc. Symposium Complex Analysis,* Canterbury, 1973.

7. W. H. J. Fuchs, Topics in Nevanlinna Theory, *Proc. NRL Conference on Classical Function Theory,* Washington, D.C., 1970.

8. W. H. J. Fuchs, A theorem on min log $|f(z)|/T(r,f)$, *Proc. Symp. Complex Analysis,* Canterbury, 1973.

9. V. P. Petrenko, The growth of meromorphic functions of finite lower order, *Izv. Ak. Nauk U.S.S.R.* **33** (1969), 414–454.

10. A. Weitsman, Asymptotic behavior of meromorphic functions with extremal deficiencies, *Trans. Amer. Math. Soc.* **140** (1969), 333–352.

Royal Institute of Technology
Stockholm, Sweden

and

University of Wisconsin–Madison
Madison, Wisconsin 53706

A LOOK AT WIMAN-VALIRON THEORY

W. H. J. Fuchs*

The original papers by Wiman on what is now generally referred to as Wiman-Valiron Theory appeared in Acta Mathematica in 1914 and 1916. The theory was developed further by Valiron, Saxer, Clunie, Kovari and others. The most recent contribution to it is a fine survey by W. K. Hayman [1], where bibliographical references are given. At various stages of the theory, general growth lemmas on increasing functions are needed. Since Professor Shah has made important contributions to the study of such lemmas, I chose Wiman-Valiron Theory as the topic of this talk. I shall present a growth lemma which allows a proof of the main results of the theory without the use of the fairly elaborate comparison functions which enter into the classical theory.

1. The basic growth lemma.

Proposition. Let A be a collection of intervals on the x-axis. Let $h(x)$ be an increasing, continuous function. If, for some $\alpha > 0$,

$$\sum_{n=1}^{M} \int_{I_n} dh < \alpha$$

for every finite selection of non-overlapping intervals $I_n \in A$, then

$$\int_E dh \leqslant 2\alpha ,$$

where E is the union of all intervals in A.

Proof. Suppose false. Then we can find a compact set $K \subset E$ such that

$$\int_K dh = 2\alpha + \delta > 2\alpha . \tag{1.1}$$

We may suppose that all $I \in A$ are open. If this is not the case to start with, replace each I by $I' \supset I$, $\int_{I'} dh < (1+\epsilon) \int_I dh$ and replace α by $\alpha(1+\epsilon)$. A finite number $I_1, I_2, \ldots, I_n$ cover K. We may suppose no $I_j = (a_j, b_j)$ is redundant, that is to say, each I_j contains points of K not contained in any other interval I_k. With suitable numbering, $a_1 < a_2 < \cdots < a_n$. (If $a_j = a_{j+1}$, then one of I_j, I_{j+1} is redundant.) Also, if I_2 is not redundant, $b_1 < b_2$. Similarly, $b_1 < b_2 < b_3 < \cdots$. Again, I_2 would be redundant if $b_1 \geqslant a_3$. Hence $b_1 < a_3$, so that I_1 and I_3 do not overlap. Repetition of this reasoning shows that $I_1, I_3, I_5, \ldots$ are non-overlapping and so are $I_2, I_4, \ldots$. Hence

$$\int_K dh \leqslant \sum \int_{I_j} dh < 2\alpha(1+\epsilon) .$$

---

* The author gratefully acknowledges support by the National Science Foundation under grant MPS 71-02727 A04.

This is a contradiction to (1.1), if ϵ is sufficiently small.

Lemma 1. *Let* $h(x)$ *be a continuous, increasing function. Let* E *be the set covered by the union of intervals* (x,x') $(x \geqslant x_0)$ *such that*

$$\int_x^{x'} dh < \int_x^{x'} \varphi(u)du\,, \tag{1.2}$$

where

$$\int_{x_0}^{\infty} \varphi(u)du < \infty\,.$$

Then

$$\int_E dh < \infty\,.$$

Proof. For any non-overlapping selection of intervals I_j with the property (1.2)

$$\sum \int_I dh < \alpha = \int_{x_0}^{\infty} \varphi(u)du\,.$$

The Lemma now follows from the Proposition.

2. The main quantities that enter the Wiman-Valiron theory are

The maximum term

$$\mu(r,f) = \mu(e^t,f) = \sup_{n \geqslant 0} |a_n| r^n = \sup_{n \geqslant 0} e^{-g_n + nt}\,.$$

The central index

$$\nu(r,f) = \text{largest } n \text{ such that } |a_n| r^n = \mu(r,f)\,.$$

The maximum modulus

$$M(r,f) = \sup_{|z|=r} |f(z)|\,.$$

Geometrical interpretation. The quantity $-g_n + nt$ is the intercept c on the negative y-axis of the line

$$y = tx - c$$

of slope t which passes through the point (n,g_n).

Therefore, if K is the smallest convex region in $x \geqslant 0$ which contains all the points (n,g_n), then $\log \mu(e^t,f)$ = intercept on the negative y-axis made by the support line of K with slope t. Two power-series with the same K have the same maximum terms.

Since $f(z)$ is entire,

$$\lim_{n\to\infty} \frac{g(n)}{n} = \infty\,.$$

Therefore, the region K is bounded by the interval $y \geqslant g_0$ of the y-axis and by a polygonal line $y = \tilde{g}(x)$. Wiman-Valiron theory restricts itself to properties of $f(z)$ which can be inferred from a knowledge of $y = \tilde{g}(x)$.

For convenience, we replace $\tilde{g}(x)$ by a function $y = g(x)$ with increasing, continuous derivative $g'(x)$ such that $g(x) - \tilde{g}(x) \to 0 \quad (x \to \infty)$. Such a function is easily constructed by rounding the corners of $y = \tilde{g}(x)$ and bending the straight line segments connecting the corners a little. The function

(2.1) $$F(z) = \sum_{n=0}^{\infty} e^{-g(n)} z^n$$

satisfies, by its construction,

(2.2) $$\mu(e^t, F) = (1 + o(1))\mu(e^t, f) \qquad (t \to \infty)$$

and

$$M(r,F) \geqslant (1 + o(1)) \sum_0^{\infty} |a_n| r^n \qquad (r \to \infty)$$
$$\geqslant M(r,f) .$$

3. **Comparison of maximum term with other terms of a power series.** We shall now prove that for most values of t the terms $e^{-g(n)+nt}$ of the series for $F(e^t)$ drop off rapidly from the maximum term.

As a function of x

$$-g(x) + tx$$

has a maximum at the (unique) root

$$x = \nu(t) = \nu$$

of

(3.1) $$g'(\nu) = t .$$

We shall call ν the central index, although it need no longer be an integer.

We apply Lemma 1 with $h(x) = g'(x)$. If ν is outside the set E of Lemma 1, then

$$g'(u) - g'(\nu) \geqslant \int_\nu^u \varphi(w)dw \qquad (u > \nu)$$

$$g'(\nu) - g'(u) \geqslant \int_u^\nu \varphi(w)dw \qquad (u < \nu)$$

By integration

(3.2) $$g(x) - g(\nu) - g'(\nu)(x - \nu) \geqslant \int_\nu^x du \int_\nu^u \varphi(w)dw$$

$$= \int_\nu^x (x - w)\varphi(w)dw \qquad (x > \nu)$$

$$(3.3)\qquad g(x)-g(\nu)-g'(\nu)(x-\nu)\geqslant\int_x^\nu du\int_u^\nu \varphi(w)dw$$

$$=\int_x^\nu (w-x)\varphi(w)dw$$

$$\geqslant \tfrac{1}{2}(\nu-x)^2\varphi(\nu)\qquad (x<\nu)$$

If we choose $\varphi(w)$ so that

$$\varphi(2w)>\tfrac{1}{4}\varphi(w)\qquad (w>w_0),$$

then, by (3.2),

$$(3.4)\qquad g(x)-g(\nu)-g'(\nu)(x-\nu)\geqslant\tfrac{1}{8}(x-\nu)^2\varphi(\nu)\qquad (2\nu\geqslant x>\nu>w_0)$$

For $x>2\nu$, by (3.2),

$$(3.5)\qquad g(x)-g(\nu)-g'(\nu)(x-\nu)\geqslant\tfrac{1}{4}\varphi(\nu)\int_\nu^{2\nu}(x-w)\ dw$$

$$=\tfrac{1}{8}(2x-3\nu)\cdot\nu\varphi(\nu)$$

$$\geqslant\tfrac{1}{8}\nu(x-\nu)\varphi(\nu)\ .$$

Putting

$$\tfrac{1}{4}\varphi(w)=\psi(w)$$

we can combine (3.3) and (3.4) into

$$g(x)-g(\nu)-g'(\nu)(x-\nu)\geqslant\tfrac{1}{2}(x-\nu)^2\psi(\nu)\qquad (x\leqslant 2\nu>2w_0)$$

or, using (3.1) and (3.5)

$$(3.6)\qquad \frac{e^{-g(n)+nt}}{e^{-g(\nu)+\nu t}}\leqslant\exp\{-\tfrac{1}{2}(n-\nu)^2\psi(\nu)\}\qquad (n\leqslant 2\nu>2w_0\ ,\ \nu\notin E)$$

$$(3.7)\qquad \frac{e^{-g(n)+nt}}{e^{-g(\nu)+\nu t}}\leqslant\exp\{-\tfrac{1}{2}\nu(n-\nu)\psi(\nu)\}\qquad (n>2\nu,\ \nu\notin E)\ .$$

As ν varies through the exceptional set E, t, connected with ν by (3.1), has a finite variation, by Lemma 1. In other words, (3.6) and (3.7) hold for all values of t which lie outside an exceptional set of finite measure.

As Hayman has shown, the classical results of Wiman-Valiron Theory follow easily from (3.6) and (3.7). In particular,

$$\sum_{|n-\nu|>k}|a_n|e^{nt}\leqslant\sum_{|n-\nu|>k}e^{-g(n)+nt}<\frac{A\mu(e^t)e^{-\frac{1}{2}\psi(\nu)k^2}}{k\psi(\nu)}\qquad (1<k\leqslant\nu)$$

and

$$M(e^t,f) < \mu(e^t)(\log \mu(e^t))^{\frac{1}{2}} (\log\log \mu(e^t))^{1+\delta}$$

hold outside an exceptional set of t of finite measure.

The geometrical core of the discussion is the fact that a convex curve $y = g(x)$ with $g(x)/x \to \infty$ as $x \to \infty$ must curve away from the tangent at x sufficiently rapidly, if x is outside an exceptional set of finite measure.

4. Functions of several variables. It is natural to ask for a generalization of Wiman-Valiron Theory to entire functions of several variables. If we try to generalize our mehtod to the case of functions of two variables

$$f(z_1,z_2) = \sum_0^\infty a_{n_1,n_2} z_1^{n_1} z_2^{n_2} ,$$

we can repeat the initial steps: we look for the boundary $y = g(x_1,x_2)$ of the smallest convex region containing the points $(n_1,n_2,-\log|a_{n_1,n_2}| = g(n_1,n_2))$ in R^3. Since $f(z)$ is entire, $g(x_1,x_2)/(x_1 + x_2) \to \infty$ $(x_1 + x_2 \to \infty)$. We need now a geometrical lemma giving a lower bound for the rate at which the surface $y = g(x_1,x_2)$ separates from its tangent plane $y = t_1x_1 + t_2x_2 - c$, if (t_1,t_2) lies outside an exceptional set. The formulation of such a lemma is complicated by the following facts:

a) For most values of $\underline{t}$ the "tangent plane" may be a support plane of K at a boundary point $(x_1,0)$ or $(0,x_2)$ and not a proper tangent plane.

b) The surface $y = g(x_1,x_2)$ may not "curve away" from the tangent plane in all directions for any proper tangent plane. Ex. $y = g(x_1 + x_2)$; every tangent plane touches the surface along a line $x_1 + x_2 = \text{const.}$

c) $g(x_1,x_2)$ may be infinite for most x_1, x_2. Ex. $f(\underline{z}) = h(z_1z_2)$.

I conclude with the hope that the reader may be able to find a suitable formulation and proof of the needed lemma.

There is some literature on the Wiman-Valiron Theory for several variables, the most successful attempt is due to A. Schumitzky in a thesis which is, unfortunately, still unpublished, although it was written 11 years ago [2].

References

1. W. K. Hayman, The local growth of power series: a survey of the Wiman-Valiron method, *Canad. Math. Bull.* 17 (1974), 317–358.

2. A. Schumitzky, Wiman-Valiron theory for entire functions of several complex variables, Cornell University Ph.D. Thesis, 1965.

Cornell University
Ithaca, New York 14853

FACTORIZATION OF MEROMORPHIC FUNCTIONS AND SOME OPEN PROBLEMS

Fred Gross

A. Introduction.

Factorization theory is basically a study of the ways in which a given meromorphic function can be written as a composition of other meromorphic functions. That is, given a meromorphic function f, we are interested in how many ways it can be expressed in the form $f = f_1 \circ f_2 \circ \cdots \circ f_n$. Numerous authors have studied various aspects of this problem and have contributed in varying degrees to the subject. A brief discussion concerning these contributions as well as a bibliography can be found in [14].

Our primary objective here is twofold. Firstly, to give a brief updated history of factorization, including some hitherto unpublished material and secondly, to identify what appear to be some of the more interesting unanswered questions on factorization.

We shall begin with some basic definitions followed by separate discussions of polynomials, rational functions and transcendental meromorphic functions.

B. Definitions.

Definition 1. A meromorphic function $h(z) = f_1 \circ f_2 \circ \cdots \circ f_n(z)$ is said to have factors $f_1, f_2, \ldots,$ and f_n provided that f_i is meromorphic for $i = 1,2,\ldots,n$. The above representation of h is said to be a *factorization* of h.

Definition 2. A meromorphic function is said to be *prime* if for every factorization of h at most one of the factors is non-bilinear.

Definition 3. A meromorphic function h is said to be *pseudoprime* if for every factorization of h, at most one of the factors is transcendental.

If h has a factorization $h = f \circ g$ and L is any bilinear function, then $h = (f \circ L) \circ (L^{-1} \circ g)$ is another factorization of h. This ambiguity can be eliminated by means of the following definition.

Definition 4. Two factorizations of a meromorphic function h(z) given by

$$h(z) = f_1 \circ f_2 \circ \cdots \circ f_n(z)$$

and

$$h(z) = g_1 \circ g_2 \circ \cdots \circ g_n(z)$$

are called *equivalent* if there exist $n-1$ bilinear functions $L_1(z), L_2(z), \ldots$ and $L_{n-1}(z)$ such that

$$f_1 = g_1 \circ L_1, \; f_2 = L_1^{-1} \circ g_2 \circ L_2, \ldots, \; f_n = L_{n-1}^{-1} \circ g_n .$$

Two factorizations of h containing bilinear factors shall be considered equivalent if they are equivalent after all the superfluous bilinear factors have been eliminated.

C. Polynomials.

The discussion of factorizing polynomials is considerably simplified by a number of easily verifiable facts.

I. There is a plentiful supply of prime polynomials. For example, all polynomials of prime degree are prime. For that matter, in a certain sense most polynomials are prime. In fact, if $P = S \circ T$ is a composite polynomial of degree n and S and T are polynomials of degrees n_1 and n_2 respectively, then $n = n_1 n_2$. Thus, the composite polynomial P belongs to an $n_1 + n_2$ parameter family, while general polynomials of degree n belong to an $n = n_1 n_2$ parameter family.

Specific examples of prime polynomials of composite degree are also plentiful. For example, for any prime integer p and any complex $A \neq 0$, the polynomial $z^p(z - A)$ is prime. When $p = 3$ we have a prime of degree 4. Observe that for $n = 4$ the parameter argument did not insure the existence of primes. Other specific examples of prime polynomials can be found by appropriately restricting the roots or coefficients of the polynomials. For example, if one of the roots of a polynomial P is sufficiently larger than the others, then P is prime. Generally speaking, given a polynomial, the problem of determining whether it is prime or composite is not a difficult one.

II. If P is a composite polynomial with factorization $P = R_1 \circ R_2$, where R_i, $i = 1,2$ are non-bilinear rational functions, then there exist nonlinear polynomials P_1 and P_2 such that $P = P_1 \circ P_2$ and these factorizations of P are equivalent.

III. Every polynomial can be factored into a finite number of prime factors.

IV. A polynomial has at most finitely many non-equivalent factorizations.

We shall prove the generalization of IV for rational functions in the next section.

The more difficult questions concerning the number of ways in which polynomials can be factored have been answered by Ritt [30].

V. Any factorizations of a given polynomial into prime polynomials contain the same number of polynomials; the degrees of the polynomials in one factorization are the same as those in the other except perhaps for the order in which they occur.

VI. Suppose that in a factorization of a polynomial $P(z)$ into prime polynomials

$$P(z) = \varphi_1 \circ \varphi_2 \circ \cdots \circ \varphi_r(z) \tag{1}$$

we have for an adjacent pair of polynomials

$$\varphi_i = \lambda_1 \circ \pi_1 \circ \lambda_2$$

and

$$\varphi_{i+1} = \lambda_2^{-1} \circ \pi_2 \circ \lambda_3 ,$$

where $\lambda_1(z)$, $\lambda_2(z)$ and $\lambda_3(z)$ are linear and where $\pi_1(z)$ and $\pi_2(z)$ of unequal degrees m and n, respectively, are of any of the following types:

(a) $\pi_1(z) = f_m(z)$ and $\pi_2(z) = f_n(z)$,

and where f_j is defined by

$$\cos ju = f_j(\cos u)$$

i.e., $f_n = 2^{n-1} T_n$, where T_n is the Chebyshev polynomial of degree n;

(b) $\pi_1(z) = z^m$ and $\pi_2(z) = z^r g(z^m)$, where $g(z)$ is any polynomial in z and r is an integer;

or

(c) $\pi_1(z) = (g(z))^n$ and $\pi_2(z) = z^n$.

Then there exists a factorization of $P(z)$ given by

$$P(z) = \varphi_1 \circ \varphi_2 \circ \cdots \circ \varphi_{i-1} \circ \theta_1 \circ \theta_2 \circ \varphi_{i+2} \circ \cdots \circ \varphi_r$$

in which $\theta_1(z)$ is of degree n and $\theta_2(z)$ is of degree m and this factorization is not equivalent to the factorization (1).

VII. If $P(z)$ has two non-equivalent factorizations, one can pass from either to a factorization equivalent to the other by repeated steps of the three types indicated in VI.

I through VII above pretty much sum up the essential facts about polynomial factorization. We now proceed to a discussion of the rational functions.

D. Rational Functions.

An analogous discussion for the factorization for rational functions is much more difficult. There is a much greater variety of possibilities. Ritt [30] claims that there are even cases in which the number of prime factors in one factorization is different from that in another. Ritt stated further that he would "return to this matter in a later communication." This author has not been able to find any further discussions of Ritt's assertion in any of Ritt's subsequent publications. Nor has the author been able to prove or disprove Ritt's statement. Thus, we have:

Conjecture 1. There exists a rational function having two factorizations into primes where the number of prime factors in one is different than the number of prime factors in the other.

Though Ritt's results may not be extendable to rationals, nevertheless, analogues of I, III and IV in C do hold for rationals.

I. If P and Q are two polynomials without common zeros and each is either linear or of prime degree, then P/Q is a prime rational. One can show further by means of a parameter argument, a little less obvious than the one we used for polynomials, that for any composite number n there exist polynomials P and Q of degree n such that P/Q is prime. On the other hand one can readily verify that there exist nonlinear polynomials P_1 and Q_1 and nonlinear polynomials P_2 and Q_2 such that with $R_i(z) = P_i(z)/Q_i(z)$ $(i = 1,2)$, the rational function $R_1 \circ R_2 = P/Q$, where P and Q are prime polynomials.

II. Though there is no uniqueness of factorization for rational functions generally, there are many rational functions other than polynomials which have uniqueness of factorization. In fact, for any polynomial P with unique factorization, P(L(z)) will be a rational function with unique factorization whenever L is bilinear. For example, if we take $P(z) = z^4 + z^2$ and $L(z) = z/(z+1)$, then

$$P(L(z)) = \frac{2z^4 + 2z^3 + z^2}{(z+1)^4}$$

has unique factorization. There are also many obvious examples of rational functions with no unique factorization. A somewhat less obvious example is

$$R(z) = \left[\frac{w^2 - 1}{w^2 - 4}\right] \circ \left[\frac{\frac{7}{4}z^2 + \frac{1}{4}}{\sqrt{2}\,z}\right]$$

$$= \left[\frac{w^2 - 3w - \frac{1}{4}}{w^2 - 3w - \frac{25}{4}}\right] \circ \left[\frac{\frac{7}{4}z^2 + \frac{3}{2}z - \frac{1}{4}}{z}\right]$$

III. Every rational function can be factored into a finite number of prime factors.

IV. A rational function has at most finitely many non-equivalent factorizations.

Proof. Let $R(z) = P/Q$ be any given rational function. We may assume, without any loss of generality, that P and Q are polynomials of the same degree and that they have no common zeros. For any factorization $R = R_1 \circ R_2$ of R, we may assume that for $i = 1,2$, $R_i = P_i/Q_i$, where P_i and Q_i have no common zeros and are polynomials of the same degree. If

$$\prod_{i=1}^{n} (w - a_i) \qquad \text{and} \qquad Q_1(w) = \prod_{i=1}^{n} (w - b_i)\,,$$

then we may write

$$\frac{P}{Q} = \frac{(P_2 - a_1 Q_2) \cdots (P_2 - a_n Q_2)}{(P_2 - b_1 Q_2) \cdots (P_2 - b_n Q_2)} \qquad (2)$$

Clearly, the numerator and the denominator on the right side have no common zeros. Thus, we must have, for some constants K_1 and K_2,

$$\prod_{i=1}^{n} (P_2 - a_i Q_2) = K_1 P \qquad \text{and} \qquad \prod_{i=1}^{n} (P_2 - b_i Q_2) = K_2 Q\,.$$

Let the zeros of P be $c_1, \ldots, c_t$. We partition the zeros of P into exactly n classes S_i $(i = 1,2, \ldots, n)$ with the i-th class having t_i elements. Generally, $t_i = t/n$, with the possible exception of one value of i. For each i, if L_i is a polynomial which maps the points in S_i into zero, then the family of polynomials with this property is the one parameter family CL_i, where C is an arbitrary complex number. If for any constant C we consider L_i and CL_i equivalent, then since the number of partitions of $\{c_1, \ldots, c_t\}$ is finite, there exist only finitely many, say k, equivalent classes which take some subset of $\{c_1, \ldots, c_t\}$ into zero.

Now suppose that $R = P/Q$ has infinitely many factorizations. Then we have infinitely many constants a_α and pairs of polynomials P_α and Q_α, $\alpha \in A$, for some index set A, such that $P_\alpha - a_\alpha Q_\alpha$ lies in one of the equivalence classes. Thus, for any $k+1$ factorizations of P/Q we may choose constants a_{k_i} $(i = 1,2)$ and a pair of rationals P_{k_1}/Q_{k_1} and P_{k_2}/Q_{k_2} such that

(3) $$P_{k_1} - a_{k_1} Q_{k_1} = L_1 \qquad L_1 \text{ a polynomial}$$

and

(4) $$P_{k_2} - a_{k_2} Q_{k_2} = \gamma L_1 \qquad \gamma \text{ a nonzero constant.}$$

Similarly, we may choose constants b_{k_i} $(i = 1,2)$ with $(a_{k_1} - b_{k_1})(a_{k_2} - b_{k_2}) \neq 0$ such that

(5) $$P_{k_1} - b_{k_1} Q_{k_1} = L_2 \qquad L_2 \text{ a polynomial}$$

and

(6) $$P_{k_2} - b_{k_2} Q_{k_2} = \lambda L_2 \qquad \lambda \text{ a nonzero constant.}$$

Solving (3) through (6) for P_{k_1}/Q_{k_1} and P_{k_2}/Q_{k_2} we obtain

$$\frac{P_{k_1}}{Q_{k_1}} = \frac{b_{k_1} L_1 - a_{k_1} L_2}{L_1 - L_2} \quad \text{and} \quad \frac{P_{k_2}}{Q_{k_2}} = \frac{b_{k_2} \gamma L_1 - a_{k_2} \lambda L_2}{\gamma L_1 - \lambda L_2} .$$

One readily verifies that one can pass from P_{k_1}/Q_{k_1} to P_{k_2}/Q_{k_2} by means of a bilinear function and the proof is complete.

The more difficult questions for rational functions remain unanswered. These answers, as we indicated earlier, if they can be found, will probably be far more complicated than V, VI and VII in C. Nevertheless, substantial results have been obtained concerning related topics such as permutable rationals, equations of the form $R \circ R_1 = R \circ R_2$ where R, R_1 and R_2 are rationals and iterates of rationals.

References on these subjects can be found in [13].

We now move on to a discussion of transcendental meromorphic functions.

E. Transcendental Functions.

We have seen that prime polynomials and prime rational functions are relatively easy to come by and that generally, deciding whether or not a given rational function is prime or composite is not a difficult problem. This does not seem to be the case for transcendental functions. Though one can easily show that for any positive integer n the function ze^{z^n} is prime, this situation is not typical. A less trivial example is the function $e^z + z$. Rosenbloom [31] stated in 1952 that this function is prime. This was subsequently proved by the author [10]. More recently, Baker and the author [3] proved that $e^z + P(z)$ is prime for any nonconstant polynomial $P(z)$. In fact, more is true. The author and Yang [19] showed that for any nonzero polynomial P and any nonconstant entire function f of order less than one, the function $Pe^z + f$ is prime. This last result is obviously not valid if e^z is replaced by e^Q, where Q is allowed to be any nonlinear polynomial.

Many other classes of primes suggested by the original example of Rosenbloom have been found. Among these are

(a) [12]. $H(z) + az$, where H is entire and periodic and satisfies, for any positive ϵ, the condition $M_H(r) < \exp(\exp(\epsilon r))$ for an infinite sequence of r, depending on ϵ and a is a nonzero constant;

(b) for any integer $n \geq 1$, $e_n(z) + z$, where $e_n(z)$ denotes the n-th iterate of e^z. The case $n = 2$ follows from (a). The more general question for $n > 2$ was posed in [12] and proved by Ozawa [25]; and

(c) [12]. $e^{H(z)+(2\pi is/\tau)z} + az$, where $H(z)$ is periodic of period τ, entire and of exponential type, s is any integer and a is any nonzero constant.

There are numerous other examples of primes of a very different form. Among these are the Gamma function [17], $\sin\sqrt{z}/\sqrt{z}$ [17] and a particularly interesting family of primes $\Sigma_{n=1}^{\infty} a_n z^{P_n}$, P_n prime and satisfying $\Sigma_{n=1}^{\infty} 1/(P_{n+1} - P_n) < \infty$. This last result was communicated to the author by I. N. Baker. Baker's method depends on results of Erdös and Macintyre [4] dealing with entire functions with gap power series and can be applied to prove the primeness of entire functions satisfying certain other gap conditions as well.

The function $\sin\sqrt{z}/\sqrt{z}$ mentioned above was one of the first examples of a transcendental entire prime of order less than 1. It turns out, in fact, that one can exhibit entire primes of arbitrary growth. This follows from the following two results, the second of which incidentally, also gives us a large family of prime functions with only real zeros.

I. [12] For any nonconstant entire periodic functions $H(z)$ and $G(z)$ of period τ and any constant λ such that $\lambda\tau/\pi i$ is irrational, the function

$$F(z) = H(z) + e^{\lambda z + G(z)}$$

is prime.

II. [12] Let $\varphi(z)$ be any entire function of lower order less than 1 or of lower order equal to 1 and lower type equal to 0. Let n_k $(k = 1,2,\ldots)$ be any infinite sequence of positive integers such that at most finitely many of the n_k are equal. Let a_k be any sequence of positive reals such that $\Pi_{k=1}^{\infty} (1 - z/a_k)^{n_k}$ is of order less than 1. If $\varphi(0) \neq \varphi(1)$, then the function

$$F(z) = e^{\varphi} z(z-1) \prod_{k=1}^{\infty} \left(1 - \frac{z}{a_k}\right)^{n_k}$$

is prime. The assertion remains valid when φ is constant.

Actually, more is true. Given an integer $k > 0$, and a constant c, $0 \leqslant c \leqslant 1$, one can construct a prime function of order k with the Nevalinna deficiency $\delta(0,f) = c$ [15]. Thus, for any $k > 0$, there exist prime entire functions of order k with prescribed Nevanlinna deficiency at 0.

These examples illustrate that transcendental entire primes can indeed be found. Though it is not at all clear, as it is in the case of polynomials and rational functions that in some sense most of the numbers in the class (entire functions in this case) are prime. Furthermore, given an arbitrary entire function it may be extremely difficult, if not impossible, to determine whether or not it is prime. For example, the following conjecture remains unsolved.

Conjecture 2 [13]. For any nonconstant polynomial Q and any nonconstant entire α, the function $F = Qe^{\alpha} + z$ is prime.

For this function as for numerous others, the question of primeness is really a question about entire functions generally.

We illustrate this by observing that Conjecture 2 above is actually equivalent to

Conjecture 2'. For any two nonlinear entire functions f and g, at least one of which is transcendental, the function $f \circ g$ must have infinitely many fix-points, i.e. $f \circ g(z) - z = 0$ has infinitely many roots.

This conjecture has a brief history. In 1920 (see [30]) Fatou proved that for any transcendental entire function f, $f \circ f$ has infinitely many fix-points. When f is of order less than ½, Fatou's result can be replaced by the more general result that $f \circ f(z) - az$ has infinitely many zeros for every $a \neq 0$.

Conjecture 3. If f is transcendental entire and a is any nonzero constant, then $f \circ f(z) - az$ has infinitely many zeros.

In 1952 Rosenbloom [31] generalized Fatou's results by proving the following.

III. Let f and g be any two nonlinear entire functions at least one of which is transcendental, then either f or $f \circ g$ has infinitely many fix-points. The case where f or g is a polynomial also follows from a more general recent result of Propkopovič [28].

The author [13] generalized Rosenbloom's result and extended it to meromorphic functions. This more general result can be stated as follows:

IV. Let $F(F^*)$ denote the family of entire (meromorphic) functions with at most a finite number of fix-points. Then (i) every entire function has at most one factorization $f \circ g$, f transcendental, $f \in F$, g entire; (ii) every meromorphic function has at most two distinct factorizations $f_i \circ g_i$, f_i meromorphic, not rational, $f_i \in F^*$, g_i entire, $(i = 1,2)$.

Some interesting consequences of this result are:

(1) Let $f(z)$ be a transcendental entire function, and let $g(z)$ be nonlinear entire. If $f \neq g$ and $f \circ g = g \circ f$, then either f or g has infinitely many fix-points.

(2) (Rosenbloom [31]) For $n > 1$, any n-th iterate $f_n(z)$ of any entire function f has infinitely many fix-points.

(3) (Rosenbloom [31]) If f and g are transcendental entire, then either f or $f \circ g$ has infinitely many fix-points.

(4) If f is transcendental meromorphic and g and h are transcendental entire, then one of the functions $f(z)$, $f \circ g(z)$ and $f \circ g \circ h(z)$ has infinitely many fix-points.

(5) If f is a nonconstant periodic entire function and g is nonconstant and entire, then $f \circ g$ has infinitely many fix-points.

(5) says that Conjecture 2′ is true when f is periodic. It is also true when either f or g is a nonlinear polynomial [31].

When $f \circ g$ is of finite order, the author and Yang [18] proved the following more general theorem which has Conjecture 2′ as an immediate consequence for this case.

V. Let $p(z)$ be a nonconstant polynomial of degree m, and let $h(z)$ and $k(z)$ be two entire functions of order less than m, where h is not identically equal to zero and k is not identically constant. Then $he^p + k$ is prime if and only if the only possible common polynomial right factors of he^p and k are linear functions.

It is natural to conjecture than an analogue of Conjecture 2′ holds for meromorphic functions as well. We will, however, refrain from making such a conjecture until more is known about the case of entire f.

The above results illustrate that information about the primeness of even specific functions may yield considerable information about interesting properties of entire and meromorphic functions generally. Though a general factorization theory for entire or meromorphic functions seems extremely difficult to achieve, it is encouraging to observe that even these seemingly rather specialized results are not as specialized as they seem but in reality reveal important properties of entire and meromorphic functions.

One can exhibit classes of primes other than those exhibited here. One can also go into a lengthy discussion of pseudoprimes. Such a discussion, though it involves rather interesting results for certain important classes of functions, would have a very similar flavor to what we have already discussed and it would be better to refer the reader who wishes to pursue this aspect further to [13], [20], [7] and [8]. The latter two contain some rather interesting methods and results of R. Goldstein. It would be more interesting at this point to move on to questions of uniqueness of factorization or more generally to the question of how many factorizations a given transcendental meromorphic function has. We would also like to demonstrate how the latter ties in with questions of periodicity and then proceed to discuss some of the interesting questions about the factorization of periodic functions.

Certainly the factorizations of transcendental meromorphic functions are far from unique. Any entire function of the form $\Sigma_{n=1}^{\infty} a_n (z^{p_1})^{p_2 n}$, where p_1 and p_2 are prime, can always be factored as either

$f_1(z^{p_1})$ or $f_2(z^{p_2})$, where $f_1(w) = \Sigma_{n=1}^{\infty} a_n w^{p_2 n}$ and $f_2(w) = \Sigma_{n=1}^{\infty} a_n w^{p_1 n}$ and where z^{p_1} and z^{p_2} are clearly prime factors.

There are also transcendental functions which have two distinct factorizations into primes. For example, $f(z) = z^2 e^{2z^2}$ has the two factorizations $(we^{2w}) \circ (z^2)$ and $(w^2) \circ (ze^{z^2})$, where all factors in both factorizations are prime. On the other hand there exist transcendental entire functions which do have unique factorization. For example, the function $z^2 e^{2z} = (w^2) \circ (ze^z)$ is one such function. H. Urabe, in a recent communication to the author, has provided a more general class of examples; the class of functions of the form $(z + P(e^z)) \circ (z + Q(e^z))$, where P and Q are nonconstant polynomials.

The general situation is even more complicated than these examples indicate. Unlike the rational functions, transcendental entire or meromorphic functions may have infinitely many factorizations. Examples of functions with infinitely many factorizations are e^z, $\cos z$ and the Weierstrass $\wp$-function. Of course, there are many others, including the rather interesting Kneser function [22], f, which has the property that for any complex number c such that $e^c = c$, $f(cz) = e^{f(z)}$. Clearly, for any positive integer n, $f(z) = e_n^{f(z/c^n)}$ where e_n denotes the n-th iterate of e^w.

These examples suggest a number of questions, all of which thus far remain unsolved.

Question 1. Does there exist a meromorphic function which cannot be factored into prime factors?

Question 2. Does there exist a meromorphic function which has no prime factors at all? The Kneser function seems like a likely candidate. It may, however, possess factorizations other than the one we exhibited, which may have prime factors.

One can also ask:

Question 3. Does there exist a meromorphic function with a nondenumerable number of nonequivalent factorizations?

Though the answer to this last question is unknown, it is known [10] that for any given meromorphic function $f(w)$ and any given meromorphic function $h(z)$, the equation

$$f \circ g(z) = h(z) \tag{7}$$

has at most denumerably many meromorphic solutions $g(z)$. When f in equation (7) is replaced by a rational function, then it is known [10] that (7) has at most finitely many meromorphic solutions $g(z)$.

A simple application of this last fact yields our first result on periodic functions.

VI. Let R be any nonconstant rational function and let g be meromorphic. Then $R \circ g$ is periodic if and only if g is.

This generalizes an earlier result of A. and C. Renyi [29] who proved VI when R is a polynomial and f is entire. It is natural to inquire if R can be replaced by a more general function. Indeed, it can be. Baker and the author proved independently (see for example [10]):

VII. Let f be entire and of order less than ½. Let g be meromorphic. Then $f \circ g$ is periodic if and only if g is.

The function $\cos z$ shows that ½ is best possible in VII.

It is reasonable to conjecture the following:

Conjecture 4. Let f be meromorphic of order less than ½. Let g be entire. Then $f \circ g$ is periodic if and only if g is.

As a consequence of IV we can also replace R in VI by any meromorphic function in F^*. We have:

VIII. Suppose that f is a function in F^* and that g is meromorphic, then $f \circ g$ is periodic if and only if g is.

Thus, for certain classes of left factors of periodic functions all right factors are ruled out except periodic ones. Are there any classes of right factors of periodic functions which are ruled out without any restrictions made on the left factors? Indeed, there are. A. and C. Renyi and Baker [1], [28] independently proved

IX. If P is a polynomial of degree greater than 2 and if f is transcendental and entire, then $F = f \circ P$ is not periodic.

This result, too, can be derived from a more general result on the solution of (7). Baker and the author proved [2]:

X. If $f(z)$ is a nonconstant entire function and $p(z)$ and $q(z)$ are nonconstant polynomials such that $f \circ p = f \circ q$, then either (i) there exist a root of unity λ and a constant β such that $p(z) = \lambda(q(z)) + \beta$ or (ii) there exist a polynomial $r(z)$ and constants c, k such that $p(z) = (r(z))^2 + k$, $q(z) = (r(z) + c)^2 + k$. In case (i), either $\lambda = 1$ and $f(z)$ is periodic with period β, or λ is a primitive j-th root of unity, $j > 1$, in which case $f(z)$ has the form

$$f(z) = \sum_{n=0}^{\infty} a_n (z - \eta)^{n_j}, \qquad \eta = \frac{\beta}{1-\lambda} .$$

If $c \neq 0$ in case (ii), then $f(k + z^2)$ is an even periodic function of period c. Clearly, all the cases mentioned above do occur.

IX follows from X by taking $q(z) = p(z + \tau)$, where τ is a period of $f \circ p$.

X suggests:

Question 4. Does there exist an analogous result when f is meromorphic?

By a method similar to the one used by Baker and the author [2], L. Flatto [5] proved the following extension of X.

XI. Let p and q be two polynomials of the same degree. Let $f \circ p = g \circ q$, where f and g are two nonconstant entire functions. Then either (i) $p(z) = \lambda q(z) + a$, where λ and a are constants or (ii) $p(z) = r^2(z) + a$ and $q(z) = br^2(z) + cr(z) + d$, where $r(z)$ is a polynomial and b, c and d are constants with $b \neq 0$.

Flatto, in a communication to the author, posed the following:

Question 5. Is there an analogue to XI when p and q are not of the same degree?

We now return to the factorization of periodic functions. Fuchs and the author [6] showed that an analogue for IX exists when f is meromorphic. Indeed, one has:

XII. Let $f(z)$ be a nonconstant meromorphic function and let $p(z)$ be a polynomial of degree n. The function

$$F(z) = f \circ p$$

cannot be periodic unless n has one of the values 1, 2, 3, 4, or 6.

If $n = 1$, then $F(z)$ can be any periodic meromorphic function. If $n = 2$, then $F(z)$ is obtained by simple changes of variable from an even periodic function. If $n \geqslant 3$, then F is an elliptic function and $F(z) = g[(z + \alpha)^n]$ for a suitable meromorphic g and complex α.

In a communication to the author in 1963, C. Renyi conjectured that XI remains valid when P is replaced by any entire function of order less than 1. This conjecture was proved by G. Halasz in 1972 [21]. We state Halasz's result explicitely.

XIII. Let f be entire and let g be entire, of order less than 1. If $f(g)$ is periodic, then g must be a polynomial of at most second degree.

XII and XIII suggest the following:

Conjecture 5. XII remains valid when P is replaced by any entire function of order less than 1.

Thus far, we have primarily discussed factorizing periodic functions. What can be said about factorization of elliptic functions? It turns out that one can say a number of interesting things. We list some of these results.

XIV. [9] Let h be an elliptic function with left and right factors f and g respectively. If g is entire, then f has no defiect values.

XV. [9] An elliptic function cannot have a periodic left factor.

XVI. [9] A right factor of an elliptic function of valence 2 which is not itself elliptic must be either a polynomial of degree 2 or of the form $A \cos(Cz + \tau) + B$, where A, B, C and τ are constants. Such factorizations in fact actually occur.

XVII. [9] Let $\wp = (1/z^2) + (g_3/28)z^4 + \cdots$ (i.e. $g_2 = 0$), where $\wp$ satisfies the differential equation

$$[h'(z)]^2 = 4h^3(z) - g_2 h(z) - g_3 .$$

Then $\wp$ is pseudoprime, and its only possible nonelliptic right factors are cubic polynomials.

XVIII. [9] A transcendental right factor of an elliptic function must be periodic, if it is entire and elliptic otherwise.

and

XIX. [9] An elliptic function η has a common right factor with a co-periodic elliptic function φ of valence 2 if and only if η is a rational function of φ.

We have been concerned above with describing the factors of periodic and elliptic functions. Are there any periodic functions which have only trivial factorizations? That is, are there any prime periodic functions? This question was posed by the author some time ago and was subsequently resolved by Mues [23]. Mues proved the following:

XX. If $\wp$ satisfies the conditions in XVII with $g_2 \neq 0$, then its derivative $\wp'$ is prime.

In a paper not yet published, Baker and Yang have shown that there exists a prime periodic entire function. They proved

XXI. The function

$$f(z) = (e^z - 1)(\exp(\exp(-z + e^z)))$$

is prime.

Recently, Ozawa [25] has also exhibited a prime entire function. He proved

XXII. The function

$$F(z) = \prod_{n=1}^{\infty}\left(1 - \frac{e^z}{\exp e^n}\right)\prod_{n=1}^{\infty}\left(1 - \frac{e^{-z}}{\exp(\exp e^n)}\right)$$

is prime.

We observe that Ozawa's function is of order 1. More generally, it is reasonable to expect the following:

Conjecture 6. Given any real number $k \geqslant 1$, there exists a prime periodic entire function of order k.

Before leaving the subject of periodic functions it would be tempting to investigate the possibility of representing periodic functions as iterates of entire functions. The only iterates of entire functions known to be periodic, however, are iterates of periodic functions. Thus, it would make little sense to pursue this further except to suggest the following:

Conjecture 7. Let f be entire and let n be a positive integer. Then the n-th iterate, f_n, of f is periodic if and only if f is.

We have seen that one can indeed say some interesting things about factorization of periodic and elliptic functions and that some of these results follow from information about the solutions of (7). That one cannot say more generally about the solutions of (7) is rather surprising. At this point all we seem to know about these solutions is that they are at most denumerable. On the other hand, for any meromorphic function f, we know of no entire solutions g_1 and g_2 of the equation

(8) $$f \circ g_1 = f \circ g_2$$

where g_1 and g_2 are not closely related in some obvious sense. This is what seems surprising. One should either be able to find solutions g_1 and g_2 that are different in an essential way or be able to show that (8) forces g_1 and g_2 to be, loosely sepaking, almost the same. The problem of finding solutions of (8), in fact, does seem to be a difficult one. Before concluding our discussion we would like to at least mention some rather interesting problems related to the latter.

For any given complex number c we consider, instead of (8)

$$f \circ g_1 - c = f \circ g_2 - c . \tag{9}$$

Let $S_c = \{w \mid f(w) - c = 0\}$. Clearly, any entire solutions g_1 and g_2 of (8) must satisfy the following condition for each c:

$$g_1^{-1}(S_c) = g_2^{-1}(S_c) .$$

That is, the set of points that g_1 maps onto S_c is equal to the set of points that g_2 maps onto S_c, with a point appearing with multiplicity n being counted n times.

What can one say about the functions g_1 and g_2 when the sets S_c are given? Consider the case when four sets are given, each with a single point, say

$$S_i = \{a_i\}, \qquad i = 1,2,3,4$$

and the a_i's all distinct. In this case, Nevanlinna [24] proved:

XXIII. Let f_1 and f_2 be two entire (meromorphic) functions, and let a_j, $j = 1,2,3,4$ ($j = 1,2,3,4,5$) be four (five) distinct complex numbers such that the equations

$$f_i - a_j = 0$$

have exactly the same roots for $i = 1,2$ for each fixed j. Then f_1 is identically equal to f_2.

That three points do not suffice for an entire function is clear from the example $g_1(z) = e^z$ and $g_2(z) = e^{-z}$.

It can be verified that for any given $n \geqslant 1$, the sets in XXIII can be replaced by sets containing n elements each.

For rational functions, one needs only 3 points.

XXIV. [13] Let f_1 and f_2 be two rational functions such that for three distinct complex numbers a_1, a_2 and a_3, the equations

$$f_i - a_j = 0$$

have exactly the same roots for $i = 1,2$ for each fixed $j = 1,2,3$ and where a point of multiplicity n is counted n times. Then f_1 is identically equal to f_2.

The multiplicity assumption is necessary. For different multiplicities A. K. Pizer [26] showed that

$$f_1(z) = \frac{4z^3}{(z-1)^3(z+1)} \quad \text{and} \quad f_2(z) = \frac{4z}{(z-1)(z+1)^3}$$

take on 0, 1 and ∞ at the same points.

More generally, for 3 finite sets, the following is known [11].

XXV. Let S_i $(i = 1,2,3)$ be distinct finite sets of complex numbers such that no set is equal to the union of the other two, and let T_i $(i = 1,2,3)$ be any finite sets of complex numbers having the same number of elements as S_i $(i = 1,2,3)$ respectively. Let $f(z)$ and $g(z)$ be two entire functions such that

$$f^{-1}(S_i) = g^{-1}(S_i) .$$

for $i = 1,2,3$. Then $g(z)$ and $g(z)$ are algebraically dependent.

For specific sets, these algebraic relationships can be quite strong and it is reasonable to expect the functions in question to be equal, provided that one chooses appropriate sets S_i.

This is fact is the case. We first state:

XXVI. [11] Let f and g be nonconstant entire functions, and let $S_1 = \{1\}$, $S_2 = \{-1\}$ and $S_3 = \{a,b\}$, where a and b are arbitrary constants such that $S_i \cap S_j = \emptyset$ for $i \neq j$. Suppose that

$$f^{-1}(S_i) = g^{-1}(S_i)$$

for $i = 1,2,3$ with the same multiplicities. Then f and g must satisfy one of the following relations:

$$f = g$$

$$fg = 1$$

or

$$(f-1)(g-1) = 4 .$$

It follows that if one chooses a and b such that $b \neq 1/a$ and $b \neq 1 + (4/(a-1))$, then $f = g$.

Thus, we see that there exist three finite sets S_i $(i = 1,2,3)$ such that any two entire functions f and g satisfying $f^{-1}(S_i) = f_g^{-1}(S_i)$ for $i = 1,2,3$ must be identical.

This suggests our next open question.

Question 6. Can one find two (or possibly even one) finite sets S_i $(i = 1,2)$ such that any two entire functions f and g satisfying $f^{-1}(S_i) = g^{-1}(S_i)$ for $i = 1,2$ must be identical?

We remark that for infinite sets one can find even single sets S, such that any two entire functions f and g satisfying $f^{-1}(S) = g^{-1}(S)$ must be identical.

The author and S. Koont have studied pairs of sets, each containing no more than two elements. In these cases one can probably prove that Question 6 can be answered negatively. If the answer to Question 6 is affirmative, it would be interesting to know how large one or both sets would have to be.

There is yet another way of looking at the problem of solving equation (8). Suppose $f \circ h = F$. For any c, let

$$T_c = \{z \mid F(z) = c\} .$$

Then $h(T_c) = S_c$. This suggests the following definition:

Definition 6. A set T is said to be a nontrivial preimage set if there exists a nonlinear entire function f and a set S containing more than one point, such that $f(T) = S$.

The problem of determining which finite sets are nontrivial preimage sets clearly, involves only nontrivial preimages of polynomials. These problems can be essentially solved. The more difficult problem is that of characterizing the infinite nontrivial preimage sets. It can be shown that there are sets both finite and infinite, which are not nontrivial preimage sets. Generally, however, a complete characterization of such sets seems to be quite difficult. Some of the more specific questions are also worth mentioning.

Question 7. Given a nontrivial preimage set S, can one add a finite number of points (possibly even one) to S such that the new set is no longer a nontrivial preimage set?

Question 8. Given a nontrivial preimage set S, is it possible to add an infinite discrete set to S, so that the new set is no longer a nontrivial preimage set?

The following fact, however, can be shown:

XXVII. For any given finite set S, one can add an infinite set T, such that $S \cap T = \emptyset$ and $S \cup T$ is not a nontrivial preimage set of any finite set.

This generalizes the following result of Rubel and Yang [31].

XXVIII. Corresponding to each infinite sequence $\{a_n\}_{n=1}^{\infty}$ in the complex plane, there exists a discrete infinite sequence $\{b_n\}_{n=1}^{\infty}$ of complex numbers disjoint from $\{a_n\}_{n=1}^{\infty}$ such that no entire function f satisfies $f(a_n) = 0$ and $f(b_n) = 1$ for $n = 1,2,3,\ldots$

F. Concluding Remarks.

The questions and conjectures we have introduced here are, of course, but a small sampling of the fairly large gaps that have yet to be filled in the study of factorization.

Some of the methods that have been used in factorization include arguments involving (a) growth, (b) value distribution, (c) conformal mapping, (d) properties of the inverse function, (e) properties of the Taylor coefficients and (f) properties of special classes of functions such as periodic functions, functions with finitely many fix points and functions which are iterates of entire functions. More information about these methods as well as others and their application to other type problems can be found in [13] and [16]. It is hoped that the reader with the help of these methods and possibly with new ones as well, will help us resolve some of our questions or prove some of our conjectures and thus provide us with further insight into the subject.

References

1. I. N. Baker, On Some Results of A. Renyi and C. Renyi Concerning Periodic Entire Functions, *Acta. Sci. Math. (Szeged)* 27 (1966), 197–200.

2. I. N. Baker and F. Gross, On Factorizing Entire Functions, *Proc. Lond. Math. Soc.* **18** (1968), 69–76.

3. ___________, Further Results on Factorization of Entire Functions, *Proc. Symp. in Pure Math. Entire Functions and Related Parts of Analysis* **11** (1968), 30–35.

4. P. Erdos and J. Macintyre, Integral Functions with Gap Power Series, *Edinburgh Math. Soc. Proc.* **10** (1954), 62–70.

5. L. Flatto, A Theorem on Level Curves of Harmonic Functions, *J. London Math. Soc.* **1** (1969), 470–472.

6. W. H. J. Fuchs and F. Gross, Generalization of a Theorem of A. and C. Renyi on Periodic Functions, *Acta. Sci. Math. (Szeged)* **32** (1971), 83–86.

7. R. Goldstein, On Factorization of Certain Entire Functions, *J. London. Math. Soc.* **2** (1970), 221–224.

8. ___________, On Factorization of Certain Entire Functions II, *Proc. Lond. Math. Soc.* **3** (1971), 267–272.

9. F. Gross, On Factorization of Elliptic Functions, *Canadian J. Math.* **20** (1968), 486–494.

10. ___________, On Factorization of Meromorphic Functions, *Trans. A.M.S.* **131** (1968), 215–222.

11. ___________, On the Distribution of Values of Meromorphic Functions, *Trans. A.M.S.* **131** (1968), 199–214.

12. ___________, Prime Entire Functions, *Trans. A.M.S.* **161** (1971), 219–233.

13. ___________, **Factorization of Meromorphic Functions,** U.S. Gov. Printing Office, 1972.

14. ___________, On Factorization Theory for Meromorphic Functions, *Comm. Math. Univ. Sancti Pauli, Tokyo* **24** (1975), 47–60.

15. F. Gross, C. Osgood and C. C. Yang, Prime Entire Functions with Prescribed Nevanlinna Deficiency, *Nagoya Math. J.* **47** (1972), 91–99.

16. ___________, On Entire Solutions of a Functional Equation in the Theory of Fluids, *J. Math. Phys.* **16** (1975), 2142–2147.

17. F. Gross and C. C. Yang, On Factorization of Entire Functions with Infinitely Many Real Zeros, *Ind. J. Pure and App. Math.* 3 (1972), 1183–1194.

18. ___________, The Fix-points and Factorization of Meromorphic Functions, *Trans. A.M.S.* **168** (1972), 211–219.

19. ___________, Further Results on Prime Entire Functions, *Trans. A.M.S.* **192** (1974), 347–355.

20. ___________, On Pseudo-prime Entire Functions, *Tohoku Math. J.* 26 (1974), 65–71.

21. G. Halasz, On Periodicity of Composed Integral Functions, *Period. Math. Hungar.* 2 (1972), 78–83.

22. H. Kneser, Reele Analytische Lösungen der Gleichung $\varphi(\varphi(x)) = e^x$ und Verwandte Funktionalgleichungen, *J. Reine Angew. Math.* **187** (1950), 56–57.

23. E. Mues, Zur Faktorisierung Elliptischer Funktionen, *Math. Z.* **120** (1971), 157–164.

24. R. Nevanlinna, **Le Théorēms de Picard-Borel et la Théorie des Fonctions Méromorphes**, Gauthier-Villars, Paris, 1929.

25. M. Ozawa, On Certain Criteria for the Left-Primeness of Entire Functions, *Kodai Math. Sem. Rep.* **26** (1975), 304–317.

26. ____________, Factorization of Entire Functions, *Tohoku Math J.* **27** (1975), 321–336.

27. A. K. Pizer, A Problem on Rational Functions, *Amer. Math. Monthly* **80** (1973), 552–553.

28. G. S. Propkopovič, The Fixed Points of Meromorphic Functions, *Ukrain. Mat. Z.* 25 (1973), 248–260.

29. A. Renyi and C. Renyi, Some Remarks on Periodic Entire Functions, *J. Analyse Math.* **14** (1965), 303–310.

30. J. F. Ritt, Prime and Composite Polynomials, *Trans. A.M.S.* 23 (1922), 51–66.

31. P. C. Rosenbloom, The Fix-points of Entire Function, *Medd. Lunds. Univ. Mat-Sem. Suppl.*, Ed. M. Riesz, (1952), 186–192.

32. A. L. Rubel and C. C. Yang, Interpolation and an Avoidable Family of Meromorphic Functions, *Mich. Math. J.* **20** (1973), 289–296.

University of Maryland, Baltimore County
Baltimore, Maryland 21228

VALUES AND GROWTH OF FUNCTIONS REGULAR IN THE UNIT DISK

W. K. Hayman

1. Introduction.

In this talk I should like to give a survey of results in the following area. Suppose that

$$w = f(z) = a_0 + a_1 z + \cdots$$

is regular in γ: $|z| < 1$ and assumes there values lying in a specified domain $D = D_f$. If $E = E_f$ is the complement of D, we also say that $f(z)$ omits the values E. What effect does this hypothesis have on the growth of $f(z)$ as measured by the maximum modulus

$$M(r,f) = \max_{|z|=r} |f(z)|, \qquad 0 < r < 1,$$

the means

$$I_\lambda(r,f) = \frac{1}{2\pi}\int_0^{2\pi} |f(re^{i\theta})|^\lambda \, d\theta, \qquad 0 < r < 1,$$

where λ is a fixed positive number, and the coefficients a_n ?

2. Theorems of Schottky and Landau.

The weakest assumption that leads to any conclusion is that E contains at least two finite values. For if $g(z)$ is an arbitrary regular function, then $f = e^g + a$ always omits the single value a and f can be chosen to grow arbitrarily rapidly. If E_f contains the values 0, 1, then Schottky's Theorem can be put [6] in the form

$$M(r,f) \leqslant \{\mu e^{\pi}\}^{\frac{1+r}{1-r}}, \qquad 0 < r < 1,$$

where $\mu = \max(1, |a_0|)$. Here the constant π is best possible at least when $a_0 = -1$. If a_0 is large or small we can obtain somewhat better results. Again Landau's theorem can be written as

$$|a_1| \leqslant 2|a_0|\{|\log |a_0|| + C\}.$$

The best upper bound for C is due to Lai [12] who proved $C \leqslant 4.76$. If $a_0 = -1$, we have $C \geqslant 4.37$ and probably this is the correct value of the constant.

If E is an arbitrary bounded set, the above results cannot be significantly sharpened. Thus, if K, α are positive constants,

$$f(z) = Ke^{\alpha\{(1+z)/(1-z)\}} = Ke^{\alpha}(1 + 2\alpha z + \cdots)$$

satisfies $|f(z)| > K$ and

$$M(r,f) = |a_0| \left\{ \frac{|a_0|}{K} \right\}^{\frac{2r}{1-r}}$$

$$a_1 = 2a_0 \{ \log |a_0| - \log K \} .$$

Thus, in order to obtain stronger results, we must assume that E_f is unbounded. With this assumption alone we can obtain a bit more, namely [7]

$$\overline{\lim_{r \to 1}} \; (1-r) \log M(r,f) = 0 ;$$

and

$$\overline{\lim_{r \to 1}} \; (1-r)^{1/3} \log M(r,f) = 0 .$$

Both these results are best possible in the sense that $(1-r)$ and $(1-r)^{1/3}$ cannot be replaced by any functions tending to zero more slowly as $r \to 1$.

3. E_f is unbounded.

We shall suppose from now on that E_f is unbounded and, in particular, that E_f contains a sequence w_n such that

(3.1) $$|w_m| \leqslant |w_{m+1}| \leqslant K w_m , \qquad 1 \leqslant m < \infty ,$$

and

(3.2) $$w_m \to \infty \quad \text{as } m \to \infty .$$

Here K is assumed to be a constant, $K > 1$. Under these hypotheses Littlewood [13] proved

(3.3) $$M(r,f) \leqslant C_1 (1-r)^{-C_2(K)} , \qquad 0 < r < 1$$

where C_1 depends on w_1, a_0 and K, but C_2 depends only on K. It can be shown [7] that the order of magnitude $C_2(K)$ satisfies

(3.4) $$C_2(K) = \frac{\log K}{2 \log \log K + O(1)} , \quad \text{as } K \to \infty .$$

Further, this result remains true if we assume that E_f contains not only the sequence w_m, but a whole sequence of disks $|w - w_m| \leqslant \alpha |w_m|$, where α is some constant, $\alpha < 1$. In this case the constant implied in $O(1)$ in (3.4) will depend on α. We shall make use of this remark later.

Let us now turn to the opposite direction, when K is nearly one. To see what to expect in this case we note that if E_f contains a connected set of values including the point w_1, then $f(z)$ is subordinate to a univalent function $F(z)$ which omits the value w_1. From these facts Littlewood [13] deduced the sharp inequality

$$M(r,f) \leqslant M(r,F) \leqslant |a_0| + \frac{4dr}{(1-r)^2} ,$$

where $d = |w_1 - a_0|$. He also obtained the right order of magnitude for $I_\lambda(r,f) \leq I_\lambda(r,F)$ and the coefficients a_n subject to these hypotheses. In particular, Littlewood proved

$$I_1(r, f - a_0) \leq I_1(r, F - a_0) \leq \frac{4dr}{1-r} .$$

Using Cauchy's inequality in the form

$$|a_n| \leq \frac{1}{r^n} I_1(r, f - a_0) , \quad \text{with } r = \frac{n}{n+1} ,$$

Littlewood deduced $|a_n| \leq 4den$. He conjectured that the factor e can be omitted here. The corresponding inequality would be sharp for every n.

Recently Baernstein [2] has taken an important step in the direction of the Littlewood conjecture by obtaining the sharp bounds for the means $I_\lambda(r,F)$ and hence $I_\lambda(r,f)$ with the above hypotheses. In particular, he proved

$$I_1(r, f - a_0) \leq \frac{4dr}{1-r^2} , \quad |a_n| \leq 2den .$$

It is natural to ask whether one can obtain corresponding results when $f(z)$ omits merely a sequence of values w_m. Let us suppose that the w_m satisfy (3.2) and

(3.5) $$\frac{w_{m+1}}{w_m} \to 1 \quad \text{as} \quad m \to \infty .$$

Then Cartwright [4] proved that

$$M(r,f) = O(1-r)^{-2+\epsilon}$$

for every positive ϵ. Here also, Baernstein and Rochberg have just (unpublished) proved the corresponding results for the means and coefficients, namely

$$I_1(r,f) = O(1-r)^{-(1+\epsilon)}, \quad |a_n| = O(n^{1+\epsilon}) .$$

Can one get rid of the ϵ? In the maximum modulus problem a complete answer is possible [8]. If E_f contains the sequence w_m satisfying (3.2) and

(3.6) $$\sum_1^\infty \left\{ \log \left| \frac{w_{m+1}}{w_m} \right| \right\}^2 < \infty ,$$

we deduce [8]

$$M(r,f) = O\{(1-r)^{-2}\}.$$

If (3.6) is not satisfied and the w_m are real and negative, there exists $f(z)$ omitting all the values w_m and such that

$$(1-r)^2 f(r) \to +\infty, \quad \text{as } r \to 1- .$$

The corresponding problem for the means and coefficients remains open, as far as I know, except under much stronger assumptions. Suppose that $f(z)$ omits one value on every circle $|w| = r$, $r > r_0$. Then the conclusions of Baernstein [2] remain valid with $d = |a_0| + r_0$. Almost at the same time, Weitsman and I [11] obtained at least the right order of magnitude under the above hypotheses, which could be weakened slightly; for instance, the conclusion $a_n = O(n)$ follows if E_f contains one value on every circle $|w| = r$, apart from a set of r of finite logarithmic measure.

To show that a countable set of omitted values w_m can be enough to imply $a_n = O(n)$, Littlewood [13] proved, by considering the modular function, that this conclusion is valid if $w_m = -m^2$. By a different method, it is possible [10] to obtain the same result for a sequence w_m of values lying on or near the negative axis and such that

$$|w_{m+1} - w_m| = O\{|w_m|^{1/2}\} .$$

While this condition could probably be weakened a little, the method depends essentially on the fact that the omitted values lie near a ray, while all the other results quoted above allowed the w_m to have arbitrary arguments.

4. The case when D_f is like a sector.

If we wish to obtain stronger results than those quoted above, we must say something about the arguments as well as the moduli of omitted values. In fact, the functions

$$F_1(z) = a_0\left(\frac{1+z}{1-z}\right)^2 = a_0(1 + \sum_{n=1}^{\infty} 4nz^n)$$

assume no value on the ray $w = Re^{i\theta}$, $0 \leqslant R < \infty$, where $\theta = -\arg a_0$. It is natural to consider similarly, the functions

$$F_\alpha(z) = a_0\left(\frac{1+z}{1-z}\right)^{2\alpha} = a_0\left(1 + \sum_1^{\infty} A(n,\alpha)z^n\right), \tag{4.1}$$

where we may take $a_0 > 0$, and $0 < \alpha < 1$. These map $|z| < 1$ onto the angle

$$S_\alpha : \arg |w| < \alpha\pi .$$

It follows from subordination that they have the largest maximum modulus and means among all functions for which $D_f \subset S_\alpha$ and a_0 is given. The coefficient problem is more complicated. If $\alpha \leqslant 1/2$, S_α is a convex domain and it then follows from a classical result of Loewner [14] that if $D_f \subset S_\alpha$ then

$$|a_n| \leqslant 4\alpha\, |a_0| .$$

The inequality is sharp for each n and α as is shown by

$$f(z) = a_0\left\{\frac{1+z^n}{1-z^n}\right\}^{2\alpha} . \tag{4.2}$$

On the other hand, if $1/2 < \alpha < 1$, the coefficients $A(n,\alpha)$ increase with n. The natural conjecture that $|a_n| \leqslant A(n,\alpha)$ in this case has been proved recently by Brannan, Clunie and Kirwan [3] and Aharonov and

Friedland [1].

We have seen that in most of the above results the means and coefficients grow less rapidly than could be deduced trivially from bounds for the maximum modulus. Taking the functions $F_\alpha(z)$ in (4.1) as our model, we have

$$M(r,F_\alpha) = |a_0|\{(1+r)/(1-r)\}^{2\alpha}$$

$$I_1(r,F_\alpha) \sim |a_0|C(1-r)^{1-2\alpha}, \quad \alpha > \tfrac{1}{2}$$

$$I_1(r,F_\alpha) \sim |a_0|C \log \frac{1}{1-r}, \quad \alpha = \tfrac{1}{2}$$

$$I_1(r,F_\alpha) \to C < \infty, \quad \text{if } \alpha < \tfrac{1}{2}.$$

Here the constant C depends on α.

From these results it is possible to deduce the correct orders of magnitude for the coefficients of subordinate functions $f(z)$, namely

$$a_n = O(n^{2\alpha-1}), \quad \alpha > \tfrac{1}{2},$$

$$a_n \to 0, \quad \alpha < \tfrac{1}{2},$$

except in the awkward case $\alpha = \tfrac{1}{2}$. Recently, Hansen [5] and Weitsman and I [11] have used an inequality of Tsuji to obtain corresponding results under more general hypotheses. The idea is embodied in the following

Theorem 1. *Suppose that* $D = D_f$ *is as above and let* $D(R)$ *be that component of* $D \cap \{|w| < R\}$ *which contains* a_0. *Let* $\omega(a_0,R)$ *be the harmonic measure of* $|w| = R$, *with respect to* $D(R)$ *at* a_0. *Then if* $2\pi r \ell(r,R)$ *is the total length of the arcs on* $|z| = r$, *where* $|f(z)| > R$, *we have*

$$\ell(r,R) \leqslant \omega(a_0,R).$$

Theorem 2. *If* $f(z)$ *is a function satisfying*

$$\ell(r,R) < (R_0/R)^{1/\beta}, \quad 0 < r < 1, \quad 0 < R < \infty, \tag{4.3}$$

then we have the following conclusions for $0 < r < 1$

$$M(r,f) < R_0\, C_1(1-r)^{-\beta}; \tag{4.4}$$

$$I_\lambda(r,f) < R_0^\lambda\, C_2, \quad \text{if } \beta\lambda < 1; \tag{4.5}$$

$$I_\lambda(r,f) < R_0^\lambda\{C_3 + \log \frac{1}{1-r}\}, \quad \beta\lambda = 1; \tag{4.6}$$

$$I_\lambda(r,f) < R_0^\lambda\, C_4(1-r)^{1-\beta\lambda}, \quad \text{if } \beta\lambda > 1, \tag{4.7}$$

and hence we have for $n \geqslant 1$

$$|a_n| < R_0 C_5, \text{ and } a_n \to 0 \text{ as } n \to \infty \text{ if } \beta < 1; \tag{4.8}$$

$$|a_n| < R_0 C_6(\log n + 1), \text{ if } \beta = 1; \tag{4.9}$$

(4.10) $$|a_n| < R_0 C_7 (n^{\beta-1}), \text{ if } \beta > 1 .$$

The constants C_j *depend only on* β *and, where relevant,* λ.

The above conclusions are all best possible, except possibly (4.9). In fact, if $a_0 > 0$ and D is the sector S_α we easily see that

$$\omega(a_0, R) \sim C(\alpha)\left(\frac{a_0}{R}\right)^{1/(2\alpha)} \qquad \text{as } R \to \infty .$$

Thus, if $D_f \subset S_\alpha$, then $f(z)$ satisfies (4.3) with $\beta = 2\alpha$. The functions $F_\alpha(z)$ given by (4.1) then show that (4.4) to (4.7) and (4.10) give the right order of magnitude. (4.2) shows that the same conclusion holds for the first inequality in (4.8).

However, the same results hold under more general circumstances. In order to illustrate the method I should like to describe a problem recently discussed by Hansen and myself (unpublished). With the above notation, suppose that D only occupies a proportion α of the open plane, in the sense that the area $A(R)$ of $D(R)$ satisfies

(4.11) $$A(R) \leqslant \pi\alpha R^2 , \quad R_1 \leqslant R < \infty$$

where α is a constant, $0 < \alpha < 1$. Our conclusion is contained in

Theorem 3. *With the hypothesis (4.11),* $f(z)$ *satisfies (4.3) and so (4.4) to (4.10) with* $\beta = \beta(\alpha)$ *where*

$$\beta(\alpha) = 2\alpha , \qquad \alpha \leqslant ½ ;$$

$$\beta(\alpha) = 1/\{2(1-\alpha)\}, \quad ½ < \alpha < 1 ,$$

and $R_0 = C_8(|a_0| + R_1)$.

The functions $F_\alpha(z)$ show that for $\alpha < ½$ the corresponding bounds for the orders of magnitude of the coefficients, means and maximum modulus are all sharp. For $\alpha = ½$, we only obtain $a_n = O(\log n)$, while the functions $F_\alpha(z)$ have bounded coefficients. It is possible that $a_n = O(1)$ under the hypotheses of Theorem 3 with $\alpha = ½$, or (perhaps even those of Theorem 2 with $\beta = 1$) but I am a little sceptical about this. At any rate, the question seems worth exploring.

The functions $F_\alpha(z)$ are certainly not extremal when α is close to one. In fact, functions $f(z)$ leaving out all the values w satisfying

$$\left|\frac{w}{K^n} - 1\right| < ½$$

for some n, such that $n = 1,2, \ldots$, certainly satisfy the hypotheses of Theorem 3 with

$$\alpha = 1 - \frac{A}{K^2} ,$$

where A is a suitable absolute constant, but, as we noted in (3.3), when K is large we can have

$$M(r,f) > C_1(1-r)^{-C_2(K)}$$

where $C_2(K)$ satisfies (3.4). Thus the correct value $\beta(\alpha)$ in Theorem 3 must satisfy

$$\beta(\alpha) \geqslant \frac{\log \frac{1}{1-\alpha}}{4 \log \log \frac{1}{1-\alpha} + O(1)} \qquad \text{as } \alpha \to 1.$$

5. Coefficients tending to zero.

It is clear that no assumption on D_f alone can imply more than

(5.1) $$a_n \to 0 .$$

For even the coefficients of a bounded function can tend to zero arbitrarily slowly. This makes it of interest to study conditions under which (5.1) holds. It follows from Theorem 3 that if $f(z)$ satisfies (4.11) with $\alpha < \frac{1}{2}$ we deduce (5.1). One can give a number of related conditions. Thus, it is enough to assume [11] that for $R \geqslant R_0$ the intersection of D with $|w| = R$ contains no arc of length greater than $2\pi\alpha R$, where $\alpha < \frac{1}{2}$.

Another condition of a rather different kind has been given by Pommerenke [15]. He has proved that if D_f does not contain arbitrarily large disks and E_f has positive capacity, then (5.1) holds. The first condition by itself implies [9]

$$a_n = O(1) ,$$

and it is natural to conjecture that it also implies (5.1). However, no condition which allows E_f to be countable, such as a lattice for instance, is known to imply (5.1).

Recently Pommerenke, Patterson and I (unpublished) have obtained such conditions, but the conclusions apply only to the superordinate functions

$$F(z) = A_0 + A_1 z + \cdots$$

which map $|z| < 1$ onto the infinite covering surface over a domain D. Thus, if the complement E of D is a lattice, we can show that

$$A_n = O(\log n)^{-\frac{1}{2}} \text{ as } n \to \infty .$$

Let $d(w)$ be the distance from any point w to E. Then, if

(5.2) $$d(w) \to 0 \quad \text{as} \quad n \to \infty ,$$

we can show at any rate that $A_n \to 0$. Unfortunately, subordination does not allow us to conclude that corresponding results hold for functions $f(z)$ omitting the values of E. It is not possible to operate with the means $I_\lambda(r,F)$. For if E has zero capacity, the functions $F(z)$ all have unbounded characteristic and so $I_\lambda(r,F)$ is also unbounded as $r \to 1$. To obtain our results we consider

$$A(r,F) = \sum_0^\infty n|a_n|^2 r^{2n-2} = \int_{|z|<r} |F'(z)|^2 \, dx\, dy .$$

If E is a lattice we can show, using deep results of Patterson, concerning the number of equivalent points under a discrete group of Moebius transformations of the unit disk, that

$$A(r,F) = C(1-r^2)^{-1} + O(1-r)^{\delta-1}, \quad \text{as } r \to 1,$$

where C, δ are positive constants. Further,

$$A(r,F) = o(1-r)^{-1}, \quad \text{as } r \to 1,$$

if (5.2) holds. From these results our conclusions follow.

References

1. D. Aharonov and S. Friedland, On an inequality connected with the coefficient conjecture for functions of bounded boundary rotation, *Ann. Acad. Sci. Fenn.*, ser. AI no. 524 (1972), 14 pp.

2. A. Baernstein II, Integral means, univalent functions and circular symmetrization, *Acta Math.* **133** (1974), 139–169.

3. D. A. Brannan, J. G. Clunie and W. E. Kirwan, On the coefficient problem for functions of bounded boundary rotation, *Ann. Acad. Sci. Fenn.*, ser. A1. Math. (1973), no. 523, 18 pp.

4. M. L. Cartwright, Some inequalities in the theory of functions, *Math. Ann.* **3** (1935), 98–118.

5. L. J. Hansen, Hardy classes and ranges of functions, *Mich. Math. J.* **17** (1970), 235–248.

6. W. K. Hayman, Some remarks on Schottky's theorem, *Proc. Cambridge Philos. Soc.* **43** (1947), 442–454.

7. W. K. Hayman, Some inequalities in the theory of functions, *Proc. Cambridge Philos. Soc.* **44** (1948), 159–178.

8. W. K. Hayman, Inequalities in the theory of functions, *Proc. London Math. Soc.* (2) **51** (1949), 450–473.

9. W. K. Hayman, Functions with values in a given domain, *Proc. Amer. Math. Soc.* **3** (1952), 428–432.

10. W. K. Hayman, Uniformly normal families, Lectures on functions of a complex variable, *University of Michigan Press,* 1955, pp. 119–212.

11. W. K. Hayman and A. Weitsman, On the coefficients and means of functions omitting values, *Math Proc. Cambridge Philos. Soc.* **77** (1975), 119–137.

12. W. T. Lai, Über den Satz von Landau, *Sci. Record N. S.* **4** (1960), 339–342.

13. J. E. Littlewood, On inequalities in the theory of functions, *Proc. London Math. Soc.* (2) **23** (1924), 481–519.

14. K. Löwner, Untersuchungen über die Verzerrung bei konformen Abbildungen des Einheitskreises $|z| < 1$, die durch Funktionen mit nicht verschwindender Ableitung geliefert werden, *Leipzig Ber.* **69** (1917), 89–106.

15. C. Pommerenke, On the growth of the coefficients of analytic functions, *J. London Math. Soc.* **5** (1972), 624–628.

Imperial College
London S. W. 7
England

SOME RECENT DEVELOPMENTS IN THE THEORY OF UNIVALENT FUNCTIONS

F.R. KEOGH

1. Notation and Definitions. D denotes the unit open disc $\{z: |z| < 1\}$ in the z-plane. A function f, analytic in a domain Δ, is said to be *univalent* in Δ if $f(z_1) = f(z_2) \Rightarrow z_1 = z_2$ whenever $z_1, z_2 \in \Delta$. S denotes the class of all functions f which are univalent in D and are subject to the normalization $f(0) = 0$, $f'(0) = 1$. For $0 \leqslant \alpha \leqslant 1$, a function f which is analytic in D and satisfies the conditions $f(0) = 0$, $f'(0) \neq 0$,

$$\operatorname{Re}\frac{z\,f'(z)}{f(z)} \geqslant \alpha \ , \quad z \in D \ ,$$

is said to be *starlike of order* α (or *starlike* in the case $\alpha = 0$). Such functions are characterized geometrically by the fact that each of them maps D conformally onto a domain which is starlike with respect to the origin and with correspondence between the origins. A function g(z), analytic in D, is said to be *convex* if it provides a conformal mapping of D onto a convex domain. A function h(z) is said to be *close-to-convex* if h(z) is analytic in D and

$$\operatorname{Re}\frac{h'(z)}{\phi'(z)} > 0 \ , \quad z \in D \ ,$$

for some $\phi(z)$ convex in D. It is well known that such a function is necessarily univalent in D.

Other notation and definitions will be introduced as the occasion arises.

2. The Pólya–Schoenberg Conjecture and Related Topics. Let f and g be functions that are analytic in D and have power series $f(z) = \sum_0^\infty a_n z^n$, $g(z) = \sum_0^\infty b_n z^n$ with respect to the origin. The Hadamard product $f * g$ of f and g is defined by $(f * g)(z) = \sum_0^\infty a_n b_n z^n$. In 1958 G. Pólya and I.J. Schoenberg ([1]) conjectured that if f and g are both convex in D then so is $f * g$. In the same paper Pólya and Schoenberg showed that their conjecture was valid in a number of special cases. In 1966 T.J. Suffridge ([2]) showed that, at any rate, under the hypothesis of the conjecture, $f * g$ is univalent. He showed, in fact, that $f * g$ was close-to-convex. In 1973 S. Ruscheweyh and T. Sheil-Small ([3]) established the truth of the Pólya-Schoenberg conjecture, that is

Theorem 1. *If* f *and* g *are both convex in* D, *then so is* $f * g$.

They deduced this result from the more general

Theorem 2. *If* φ *is convex and* f *close-to-convex in* D, *then* $\varphi * f$ *is close-to-convex in* D.

The proof of these theorems requires a careful study of certain geometric properties of convex and starlike functions and the expression of these in analytic form. The methods of the authors enable them to obtain a number of similar results and to settle two other conjectures. Similar to Theorem 1, particularly in view of the well-known fact that a convex function (mapping the origin onto the origin) is necessarily starlike of order ½, is

Theorem 3. *If* φ *and* ψ *are starlike of order* ½ *in* D, *then so is* $\varphi * \psi$.

This has the following equivalent form.

Theorem 3'. *If* φ *and* ψ *are odd and starlike in* D, *then so is* $\varphi * \psi$.

If f and g are analytic in D and $g(z) = f(\omega(z))$, where $\omega(z)$ is analytic and satisfies $|\omega(z)| \leq |z|$ in D, then g is said to be *subordinate* to f in D, and we write $g \prec f$. In the case when f is univalent in D, this is the analytic equivalent of the statement $g(D) \subset f(D)$, $g(0) = f(0)$. The following result was first conjectured by M.S. Wilf ([4]), and finally proved in [3].

Theorem 4. *Let* φ *and* ψ *be convex in* D *and suppose that* $f \prec \psi$. *Then* $\varphi * f \prec \varphi * \psi$.

Theorem 1 follows from Theorem 4.

If $f(z) = \sum_1^\infty a_n z^n$ is analytic in D, then the de la Vallée Poussin means of f are defined by

$$V_n(z, f) = \binom{2n}{n}^{-1} \sum_{k=1}^{n} \binom{2n}{n+k} a_k z^k, \quad n = 1, 2, \ldots .$$

In [1] it was shown that all the $V_n(z, f)$ are convex (starlike) whenever f is convex (starlike), and conversely. In particular, with $f(z) = z(1-z)^{-1}$, the polynomials

$$V_n(z) = \binom{2n}{n}^{-1} \sum_{k=1}^{n} \binom{2n}{n+k} z^k$$

are convex, and by Theorem 1 this weaker result already implies the full one. Also, by Theorem 2, it implies that all the $V_n(z, f)$ are close-to-convex whenever f is close-to-convex. In [1] it was further proved that $V_n(z, f) \prec f(z)$ for every convex f, and it was conjectured that

$$V_n(z, f) \prec V_{n+1}(z, f) \quad (n = 1, 2, \ldots) .$$

The truth of the conjecture follows immediately from Theorem 4 if it holds for $f(z) = z(1-z)^{-1}$. The latter was verified by Pólya and Schoenberg, though no proof was given in [1].

Generalizations of some of the results in this section are given in Section 7.

3. The Bieberbach Conjecture and Related Topics. Suppose that $f(z) = z + \sum_2^\infty a_n z^n \in S$. The Bieberbach conjecture is that $|a_n| \leq n$ for all n, but so far this has only been shown to be true for n = 2 ([5]), 3([6]), 4([7]), 5([8]), and 6([9], [22]). A rough estimate for $|a_n|$, namely, $|a_n| \leq 1 \cdot 243n$ for all n was obtained by I. Milin ([10]) in 1965. An earlier such estimate was $|a_n| \leq en$, due to J.E. Littlewood (see, for example [11], Theorem 248). In 1971 Milin's result was improved by C.H. FitzGerald ([12]) to $|a_n| < \sqrt{\frac{7}{6}}\, n < 1 \cdot 081n$ for all n. This was deduced from the inequalities

$$\sum_{1}^{n} k|a_k|^2 + \sum_{n+1}^{2n} (2n-k)|a_k|^2 \geqslant |a_n|^4, \quad n = 2, 3, \ldots . \tag{1}$$

(1) was obtained through a method involving, roughly speaking, 'exponentiation' of the Grunsky inequalities ([13]). In the same paper [12] by Fitzgerald a new proof is given of the well known fact that $|a_n| \leqslant n$ for all n in the case when all the a_n are real, and also a proof that $\limsup_{n\to\infty} \frac{|a_n|}{n} < 1$ except when f is of the form $f(z) = z(1-e^{i\lambda}z)^{-2}$, where λ is real. This is a weaker version of the result

$$\lim_{n\to\infty} \frac{|a_n|}{n} < 1 \quad (f(z) \neq z(1-e^{i\lambda}z)^{-2}) \tag{2}$$

obtained by W.K. Hayman ([14]).

Fitzgerald's inequality $|a_n| < 1 \cdot 081\, n$ was further improved in 1976 by D. Horowitz ([15]) to

$$|a_n| \leqslant (209/140)^{1/6} n < 1 \cdot 0691 n \tag{3}$$

for all n, using the following stronger form of (1), again due to Fitzgerald ([12]). With f as defined above, if $\lambda_1, \lambda_2, \ldots, \lambda_L$ are complex numbers and $n_1 \leqslant n_2 \leqslant \ldots \leqslant n_L$ are positive integers, then

$$\begin{aligned} \Big|\sum_{j=1}^{L} \lambda_j |a_{n_j}|^2\Big|^2 &\leqslant \sum_{j=1}^{L} |\lambda_j|^2 \Big\{ \sum_{k=1}^{n_j} k|a_k|^2 + \sum_{k=n_j+1}^{2n_j} (2n_j-k)|a_k|^2 \Big\} \\ &+ 2\,\mathrm{Re} \sum_{1\leqslant j_1<j_2\leqslant L} \lambda_{j_1}\bar{\lambda}_{j_2} \Big\{ \sum_{k=n_{j_2}-n_{j_1}}^{n_{j_2}} (n_{j_1}-n_{j_2}+k)|a_k|^2 + \sum_{k=n_{j_2}+1}^{n_{j_1}+n_{j_2}} (n_{j_1}+n_{j_2}-k)|a_k|^2 \Big\} \end{aligned} \tag{4}$$

(1) is obtained from (4) with $L = 1$. Horowitz has recently informed the author that he has obtained a slight improvement (unpublished) of (3) to

$$|a_n| \leqslant (1,659,164,137/681,080,400)^{1/14} n \cong 1 \cdot 0657 n.$$

The inequalities (1) were obtained by Fitzgerald ([12]) from his 'exponentiated' form of Goluzin's version ([16]) of the Grunsky inequalities. We state this exponentiated form as

Theorem 5. *If* $z_1, z_2, \ldots$ *are points in* D *and* $f \in S$, *then for each positive integer* n *and set of complex numbers* $\beta_1, \beta_2, \ldots \beta_n$,

$$\sum_{\mu,\nu=1}^{n} \beta_\nu \bar{\beta}_\mu \left| \frac{f(z_\nu)-f(z_\mu)}{z_\nu - z_\mu} \; \frac{1}{1-z_\nu \bar{z}_\mu} \right| \geqslant \left| \sum_{\nu=1}^{n} \beta_\nu \left| \frac{f(z_\nu)}{z_\nu} \right| \right|^2,$$

and (5)

$$\sum_{\mu,\nu=1}^{n} \beta_\nu \bar{\beta}_\mu \left| \frac{f(z_\nu)-f(z_\mu)}{z_\nu - z_\mu} \; \frac{1}{1-z_\nu \bar{z}_\mu} \right|^2 \geqslant \left| \sum_{\nu=1}^{n} \beta_\nu \left| \frac{f(z_\nu)}{z_\nu} \right|^2 \right|^2,$$

where $\frac{f(z_\nu)-f(z_\mu)}{z_\nu - z_\mu}$ *is interpreted as* $f'(z_\nu)$ *if* $z_\nu = z_\mu$.

Using only (5), Horowitz ([17]) has obtained a new proof of Hayman's result (2). He points out that his proof applies, in fact, to a class of analytic functions larger than S.

Suppose now, that $p(z) = z + \sum_{1}^{m} a_k z^k$ is a polynomial in S of degree not greater than 27. Horowitz ([18]) has also shown that $|a_k| \leqslant k$, $2 \leqslant k \leqslant 27$. The method of proof is based on the interplay of (1), (4), and the following weaker form of a lemma of J. Dieudonné [19].

Lemma 1. *If* $z + \sum_{1}^{m} a_k z^k$ *is in* S, *then*

$$|a_{m-k}| \leqslant [1 + (2|a_2|)^2 + (3|a_3|)^2 + \ldots + ((k+1)|a_{k+1}|)^2]^{\frac{1}{2}}(m-k)^{-1}$$

for $0 \leqslant k \leqslant m-1$.

Loewner's classical differential equation ([6]) provides explicit representations for the coefficients a_n of a class of univalent functions which is dense in S. The only variable element in the expression $\psi_n(K(t))$ for a_n is a function $K(t)$ of modulus 1. Loewner showed that $|\psi_3(K)| \leqslant 3$, but the functionals $\psi_n(K)$ become more and more formidable as n increases, and the subsequent proofs of the validity of the Bieberbach conjecture for n = 4, 5 and 6 ([7], [8], [9], [22]) were based on different methods. Z. Nehari ([20]) has demonstrated the possibility of a direct attack on the problem of finding $\max|\psi_n(K)|$, and has illustrated his method in the case n = 4. Though the latter case has been dealt with already by a simple, elementary argument ([21]), Nehari's basic procedure is applicable to coefficients of higher order. With the notation

$$A_m = \int_0^\infty K^m(t)e^{-mt}\,dt\,,\quad B = \int_0^\infty K^2(t)e^{-2t}\int_0^t K(s)e^{-s}ds,$$

Loewner's expression for a_4 is

$$a_4 = -8A_1^3 + 8A_1A_2 + 4B - 2A_3\,,$$

and it was by working with this expression that Nehari proved by elementary methods that $|a_4| \leqslant 4$. He suggests in [20] that, in spite of technical difficulties, the case n = 5 might be worth trying, particularly in view of the complexity of the proof of $|a_5| \leqslant 5$ by Pederson and Schiffer ([8]).

4. Integral means of univalent functions. Let $u(z)$ be subharmonic in the annulus $r_1 < |z| < r_2$ $(0 \leqslant r_1 < r_2 \leqslant \infty)$, and define

$$u^*(re^{i\theta}) = \sup_E \int_E u(re^{i\psi})d\psi \quad (0 \leqslant \theta \leqslant \pi)\,, \tag{6}$$

where the sup is taken over all measurable sets $E \subset [0, 2\pi)$ with Lebesque measure 2θ. This function was introduced by A. Baernstein in [23]. From results established there Baernstein ([24]) obtained the following.

Theorem 6. u^* *is subharmonic in the semi-annulus* $\{z\colon r_1 < |z| < r_2,\ 0 < \arg z < \pi\}$.

Theorem 6 is the main ingredient of Baernstein's proofs of the next two theorems ([24]), in which $k(z) = z(1-z)^{-2}$ denotes the Koebe function.

Theorem 7. *Let* $A(s)$ *be a convex non-decreasing function of* $\log s$. *Then for* $f \in S$ *and* $0 < r < 1$,

$$\int_0^{2\pi} A(|f(re^{i\theta})|)d\theta \leq \int_0^{2\pi} A(|k(re^{i\theta})|)\, d\theta .$$

In particular, for $0 < r < 1$,

$$\int_0^{2\pi} |f(re^{i\theta})|^p\, d\theta \leq \int_0^{2\pi} |k(re^{i\theta})|^p d\theta \ (0 < p < \infty), \tag{7}$$

$$\int_0^{2\pi} \log^+ |f(re^{i\theta})|\, d\theta \leq \int_0^{2\pi} \log^+ |k(re^{i\theta})|\, d\theta .$$

The best previously known result of the type (7) seems to be due to I.E. Bazilevič ([25]), who showed that (7) holds for $p = 1, 2$ with an appropriate universal constant added to the right hand side. It may be worth remarking that the well known distortion inequality $\max_\theta |f(re^{i\theta})| \leq r(1-r)^{-2}$ follows from (7) on taking the pth root of both sides and allowing $p \to \infty$.

Let $f(z)$ now be any analytic function univalent in D. Let Δ be the conformal image of f, and let Δ^* be the circularly symmetrised domain of Δ. Let g be the conformal map of D onto Δ^* with $g(0) = |f(0)|$.

Theorem 8. *With* f *and* g *as just described, the conclusion of Theorem 1 holds.*

An outline of the proofs of Theorems 7 and 8 is as follows. Let $h(z)$ be used to denote either $k(z)$ in Theorem 7 or $g(z)$ in Theorem 8. Let F, H be the inverse functions of f, h, and define $u = -\log|F|$, $v = -\log|H|$. Extend u, v to the whole plane by setting them equal to zero outside the images of f, h respectively. Then u is subharmonic in the plane except for a logarithmic singularity at $f(0)$, and a similar statement holds for v.

It can be shown that Theorems 7 and 8 follow from the inequality

$$\int_0^{2\pi} [u(re^{i\theta}) - \alpha]^+ d\theta \leq \int_0^{2\pi} [v(re^{i\theta}) - \alpha]^+ d\theta \ (0 < r < \infty, 0 < \alpha < \infty),$$

and this inequality is easily seen to be implied by

$$u^*(re^{i\theta}) \leq v^*(re^{i\theta}) \quad (0 < r < \infty, 0 < \theta < \pi), \tag{8}$$

where u^*, v^* are defined as in (6). Under the hypotheses of Theorem 6, the key to proving (8) is the fact that v^* is harmonic in the upper half plane, whereas, by Theorem (6), u^* is subharmonic there. Similarly, under the conditions of Theorem 8, $v^* + 2\pi \log^+ \frac{r}{|f(0)|}$ (or v^*, if $f(0) = 0$) turns out to be harmonic in the part of Δ^* that lies in the upper half-plane.

With regard to the integral means of the derivatives of univalent functions, Jinfu Feng and T.H. MacGregor ([26]) have recently proved the following result.

Theorem 9. *There exist positive constants* $D_{n,\lambda}$ *such that if* $f \in S$ *and* $\lambda > \frac{2}{5}$, *then*

$$\frac{1}{2\pi}\int_0^{2\pi} |f^{(n)}(re^{i\theta})|^{\lambda}\, d\theta \leqslant D_{n,\lambda}(1-r)^{1-(n+2)\lambda},$$

where $0<r<1$, $n=1,2,\ldots$. *If* $0<\lambda\leqslant\frac{2}{5}$, *then for each* $\epsilon>0$,

$$\frac{1}{2\pi}\int_0^{2\pi} |f^{(n)}(re^{i\theta})|^{2}\, d\theta \leqslant D_{n,\lambda}(1-r)^{(\frac{1}{2}-n)\lambda-\epsilon},$$

where the $D_{n,\lambda}$ *are constants depending only on* λ,ϵ *and* $n=1,2,\ldots$.

5. Functions of bounded boundary rotation. For $k\geqslant 2$ let V_k denote the class of functions $f(z)=z+\sum_2^{\infty} a_n z^n$ that are locally univalent in D and map D onto a domain with boundary rotation at most $k\pi$ (see [27] for the definition and basic properties of the class V_k). Analytically, this is equivalent to the condition

$$\int_0^{2\pi} \left| \mathrm{Re}\left[1+re^{i\theta}\frac{f''(re^{i\theta})}{f'(re^{i\theta})}\right]\right| d\theta \leqslant k\pi, \quad 0\leqslant r<1 .$$

V_2 coincides with the class of all convex functions, and V. Paatero ([28]) showed that all functions in V_4 are also in S. It is easily shown, through counterexamples, however, that for $k>4$ there are functions in V_k that are not in S.

The function

$$f_k(z)=\frac{1}{k}\left[\left(\frac{1+z}{1-z}\right)^{k/2}-1\right]=\sum_1^{\infty} A_n z^n$$

belongs to V_k, and the 'coefficient conjecture' for V_k was that for $f\in V_k$,

$$|a_n|\leqslant A_n \quad (n>1) . \tag{9}$$

This was proved for $n=2$ by Pick (see [27]), for $n=3$ by O. Lehto ([27]), and for $n=4$ in [29], [30], [31] and [32]. In support of the conjecture, J. Noonan [33] showed that for a given function $f(z)$ in V_k, $\lim_{n\to\infty}\frac{|a_n|}{A_n}$ exists and is less than 1 unless $f(z)=e^{-i\theta}f_k(e^{i\theta}z)$, $0\leqslant\theta\leqslant 2\pi$.

Let now $g(z)=\sum_0^{\infty} b_n z^n$ and $h(z)=\sum_0^{\infty} c_n z^n$ be analytic in D, where $c_n\geqslant 0$ for all n. If $|b_n|\leqslant c_n$ for all n then we write $g(z)<<h(z)$ (and we say that $h(z)$ majorizes $g(z)$). In [34], D.A. Brannan, J.C. Clunie and W.E. Kirwan reduced the problem of proving (9) to showing that

$$\left(\frac{1+xz}{1-z}\right)^{\alpha} << \left(\frac{1+z}{1-z}\right)^{\alpha} \tag{10}$$

for $\alpha\geqslant 1$, $|x|=1$. In the same paper the authors showed that $V_k\subset K(\beta)$, $\beta=\frac{1}{2}k-1$, where $K(\beta)$, $\beta>0$, denotes the class of 'close-to-convex functions of order β' of the form $f(z)=z+\sum_2^{\infty}a_n z^n$, analytic in D, for which there exists a starlike function $s(z)=z+\sum_2^{\infty}b_n z^n$ in D and a constant a such that

$$\left|\arg\frac{z\,f'(z)}{a\,s(z)}\right|<\frac{\beta\pi}{2}$$

in D. Among other things, in [34] the authors proved (10) for $\alpha\geqslant 2$, so establishing the corresponding coefficient conjecture for $K(\beta)$ when $\beta \geqslant 1$, and hence (9) for V_k when $k \geqslant 4$. In [35] D. Aharonov and S. Friedland established (10) completely, that is, for all $\alpha \geqslant 1$, so proving (9) for all $k \geqslant 2$ and the corresponding coefficient conjecture for $K(\beta)$ for all $\beta > 0$. More recently D.A. Brannan ([36]) has given a shorter proof of (10).

6. Asymptotic behaviour of coefficients of a function in S as a Tauberian remainder theorem. Let $f(z) = z + \sum_1^\infty a_n z^n \in S$. Hayman ([14]) proved that

$$\lim_{n\to\infty} \frac{|a_n|}{n} = \alpha \leqslant 1, \tag{11}$$

with equality only for the Koebe functions

$$e^{-i\lambda} k(e^{i\lambda} z) = z(1 - e^{i\lambda} z)^{-2} = \sum_1^\infty n e^{(n-1)\lambda i} z^n ,$$

where λ is real. Hayman's proof begins with the elementary observation that each $f \in S$ has a direction $e^{i\theta_0}$ of maximal growth, in the sense that

$$\lim_{r\to 1} (1 - r)^2 |f(re^{i\theta_0})| = \alpha \tag{12}$$

and the limit is 0 for every other direction. In particular, if $\alpha > 0$ then the direction $e^{i\theta_0}$ is unique. The second step is the deduction of (11) from (12). Hayman's argument is relatively simple for $\alpha = 0$, but quite complicated for $\alpha > 0$.

Milin ([37], [38]) simplified this difficult last step when $\alpha > 0$. His argument is essentially based on the following result which, as P. Duren ([39]) has pointed out, may be regarded as a new Tauberian theorem. Let

$$g(r) = \sum_0^\infty b_n r^n , \quad b_0 = 1 ,$$

be a power series with complex coefficients, convergent for $-1 < r < 1$. Let

$$s_n = \sum_0^n b_k , \quad \sigma_n = \frac{1}{n+1} \sum_0^n s_k ,$$

and

$$\log g(r) = \sum_1^\infty c_n r^n .$$

Theorem 10. *Suppose that* $|g(r)| \to \alpha$ *as* $r \to 1$, *and*

$$\sum_1^\infty n |c_n|^2 < \infty .$$

Then $|s_n| \to \alpha$ *and* $|\sigma_n| \to \alpha$ *as* $n \to \infty$.

This theorem is applied to the function

$$g(r) = \frac{(1-r)^2}{r} f(r) ,$$

where it is assumed, without loss of generality, that $\theta_0 = 0$. A simple calculation gives

$$s_n = a_{n+1} - a_n , \quad \sigma_{n-1} = \frac{a_n}{n} .$$

The Tauberian condition is $\sum_1^\infty n|c_n|^2 < \infty$, and this is a consequence (for $\alpha > 0$) of the following theorem of Bazilevič ([40], [38]).

Theorem 11. *Let* $f \in S$, *and let*

$$\log \frac{f(z)}{z} = 2 \sum_1^\infty \gamma_n z^n .$$

Suppose that $\alpha > 0$ *and* $\theta_0 = 0$. *Then*

$$\sum_1^\infty n \left| \gamma_n - \frac{1}{n} \right|^2 \leqslant \tfrac{1}{2} \log \frac{1}{\alpha} .$$

A step in Milin's proof of Theorem 10 is to show that the s_n are bounded. If $g(r) \to \alpha$, rather than $|g(r)| \to \alpha$, then a classical Tauberian theorem ([41, p. 154]) yields $\sigma_n \to \alpha$.

In [39] P. L. Duren shows that, under a relatively mild assumption on the behaviour of f along its ray of maximal growth, an estimate of the rate of convergence of a_n/n can be obtained (under a much stronger hypothesis, Bazilevič ([42], [40], [38]) estimated the rate of convergence of $|a_n|/n$ to α).

Theorem 12. *For some positive constants* B *and* δ, *and for some complex number* $s \neq 0$, *let* $f \in S$ *satisfy the inequality*

$$\left| \frac{(1-r)^2}{r} f(r) - s \right| \leqslant B(1-r)^\delta , \quad 0 < r < 1 .$$

Then $\left| \frac{a_n}{n} - s \right| \leqslant \frac{C}{\log n}$, $n = 2, 3, \ldots$, *where* C *depends only on* $|s|$, B, *and* δ.

Duren's proof of this theorem depends on the following Tauberian remainder theorem due essentially to G. Freud ([43]) and J. Korevaar ([44]). The notation is as in the second paragraph of this section.

Theorem 13. *For some positive constants* B *and* δ, *suppose that*

$$|g(r) - s| \leqslant B(1-r)^\delta , \quad 0 < r < 1 ,$$

and that $|s_n| \leqslant M$, $n = 1, 2, \ldots$. *Then*

$$|\sigma_n - s| \leqslant C/\log n , \quad n = 2, 3, \ldots ,$$

where C *depends only on* B, M, *and* δ.

Finally, Duren points out that a result similar to Theorem 12 holds in the case $\alpha = 0$.

7. Recent Developments in the Theory of Starlike Functions. Let $P_n(\theta)$ denote the class of polynomials

$$P(z) = \prod_{j=1}^{n} (1 + z e^{i\alpha_j}) ,$$

where $0 \leq 2\theta \leq \alpha_{j+1} - \alpha_j$, $1 \leq j \leq n$, $\alpha_{n+1} = \alpha_1 + 2\pi$. Let

$$C(n, k, \theta) = \frac{\sin n\theta \sin(n-1)\theta \ldots \sin(n-k+1)\theta}{\sin\theta \sin 2\theta \ldots \sin k\theta} ,$$

and let

$$\{A_k\} = \sum_{k=0}^{n} A_k C(n, k, \theta) z^k .$$

In [45] T.J. Suffridge recently proved a number of interesting results, among them the following.

Theorem 14. *Let* $\alpha \leq 1$ *be fixed. Then* $f(z) = z + \sum_2^\infty a_n z^n$ *is starlike of order* α *if and only if there exists a sequence* $\{P_{n_k}\}_{k=1}^\infty$, $P_{n_k} \in P_{n_k}(\pi/(n_k + 2 - 2\alpha))$, *such that* $n_k \to \infty$ *and* $zP_{n_k} \to f(z)$ *uniformly on compact subsets of* $|z| < 1$.

A similar result ([46]) was proved earlier by Suffridge for the whole class S of normalized univalent functions, working with a class of characterizing polynomials different from the P_n.

Theorem 15. *If* $\{A_k\}$, $\{B_k\} \in P_n(\theta)$ *then* $\{A_k B_k\} \in P_n(\theta)$.

Theorem 16. *If* $0 \leq \theta_1 \leq \theta_2 \leq \frac{\pi}{n}$ *and* $\{A_k\} \in P_n(\theta_1)$, *then* $\{A_k\} \in P_n(\theta_2)$. *Further,* $P_n(\theta) \subset \mathrm{co}\{Q_n^{(k)}(z;\theta)\}_{k=1}^n$.

Here co G denotes the convex hull of G. Let $C_k(\alpha)$ be defined by

$$z(1-z)^{2(\alpha-1)} = \sum_0^\infty C_k(\alpha) z^{k+1} ,$$

so that

$$C_k(\alpha) = \frac{\Gamma(2 + k - 2\alpha)}{\Gamma(2 - 2\alpha)\Gamma(k + 1)} ,$$

and with $f(z) = \sum_0^\infty C_k(\alpha) a_k z^{k+1}$, $g(z) = \sum_0^\infty C_k(\alpha) b_k z^{k+1}$, define

$$(f * g)(z) = \sum_0^\infty C_k(\alpha)\, a_k\, b_k\, z^{k+1}$$

(Note: $f * g$ here is not the Hadamard product referred to in Section 2). From Theorem 15 is deduced

Theorem 17. *If* $f(z) = \sum_0^\infty C_k(\alpha)\, a_k z^{k+1}$, $g(z) = \sum_0^\infty C_k(\alpha)\, b_k z^{k+1}$ *are starlike of order* α, *then so is*

$$(f * g)(z) = \sum_0^\infty C_k(\alpha)\, a_k\, b_k\, z^{k+1} .$$

The cases $\alpha = 0$, $\alpha = ½$ are Theorems 1, 3, respectively, of Ruscheweyh and Sheil-Small described in Section 1.

An easy consequence of Theorem 16 is

Theorem 18. *If* $\sum_0^\infty C_k(\alpha)\, a_k z^{k+1}$ *is starlike of order* α *and* $\alpha \leqslant \beta \leqslant 1$, *then* $\sum_0^\infty C_k(\beta)\, a_k z^{k+1}$ *is starlike of order* β.

When $\alpha = 0, \beta = \tfrac{1}{2}$, this reduces to the well known fact ([47], [48]) that if $\sum_1^\infty k\, a_k z^k$ is starlike then $\sum_1^\infty a_k z^k$ is starlike of order ½.

Theorem 19. *With the hypotheses of Theorem 17, suppose further that there are arcs* Γ_f, Γ_g *on* $|z| = 1$ *of length* γ_f, γ_g, *respectively, such that* f, g *are analytic and satisfy* $\mathrm{Re}\{z f'(z)/f(z)\} = \alpha$, $\mathrm{Re}\{z g'(z)/g(z)\} = \alpha$ *on* Γ_f, Γ_g, *respectively, where* $\gamma_f + \gamma_g > 2\pi$. *Then there is an arc of length* $\gamma_f + \gamma_g - 2\pi$ *on which* $(f * g)(z) = \sum_0^\infty C_k(\alpha) a_k b_k z^{k+1}$ *is analytic and satisfies* $\mathrm{Re}\{z(f * g)'(z)/(f * g)(z)\} = \alpha$.

In [49] S. Ruscheweyh defined C_α, $\alpha \leqslant 1$, to be the class of functions f analytic in D for which there exists a starlike function $g \in S_\alpha$ (the class of starlike functions of order α) such that

$$\mathrm{Re}\, \frac{z f'(z)}{g(z)} > 0\,, \quad z \in D.$$

He also defined the class R_α, $\alpha \leqslant 1$, of *prestarlike* functions of order α to consist of those functions f analytic in D such that f(0), $f'(0) \neq 0$, and

$$\begin{cases} \mathrm{Re}\, \dfrac{f(z)}{z f'(0)} > \tfrac{1}{2}\,, \quad z \in D\,, \quad \text{for } \alpha = 1\,, \\[2ex] z(1-z)^{2(\alpha-1)} * f(z) \in S_\alpha\,, \text{for } \alpha < 1\,, \end{cases}$$

where $*$ denotes Hadamard product. With this notation Ruscheweyh states the following theorems.

Theorem 20. *For* $f, g \in R_\alpha$, $\alpha \leqslant 1$, *we have* $f * g \in R_\alpha$.

Theorem 21. *For* $\alpha \leqslant \beta \leqslant 1$ *we have* $R_\alpha \subset R_\beta$.

Theorem 22. *For* $f \in C_\alpha$, $g \in R_\alpha$, $\alpha \leqslant 1$, *we have* $f * g \in C_\alpha$.

Theorem 23. *For* $f \in R_\alpha$, $\alpha \leqslant \tfrac{1}{2}$, *and every* $z_0 \in D$, *we have*

$$z\, \frac{f(z) - f(z_0)}{z - z_0} \in R_{\alpha + 1/2}\,.$$

He deduces all these results from the following main theorem.

Theorem 24. *Let* A *be the class of functions* f, *analytic in* D, *such that* $f(0) = 0$, $f'(0) \neq 0$, *and for* $\alpha \leqslant \beta \leqslant 1$, *let* $p(z) \in S_{1+\alpha-\beta}$ *be analytic in* $\bar{D}$. *With* $g \in R_\beta$, *let* T *be the continuous linear operator*

$$f \mapsto (Tf)(z) := \left[g(yz)\, \frac{p(y)}{y} *_y f(y) \right] \Big|_{y=1}$$

acting on A. *Then* $T\colon R_\alpha \to R_\beta$.

Here $*_y$ indicates that the Hadamard product is taken with respect to power series in y.

Theorems 20, 21, 22 are, respectively, slightly extended versions of Suffridge's Theorems 17, 18, and Ruscheweyh's

Theorem 25. *If* $\alpha \leq 1$ *and*

$$f(z) = \sum_0^\infty C_k(\alpha)\, a_k z^{k+1} \in C_\alpha\ ,\ g(z) = \sum_0^\infty C_k(\alpha)\, b_k z^{k+1} \in S_\alpha\ ,$$

then

$$h(z) = \sum_0^\infty C_k(\alpha)\, a_k\, b_k\, z^{k+1} \in C_\alpha\ .$$

Theorem 23 contains the following results of earlier papers of Ruscheweyh and Sheil-Small ([3]) and Suffridge ([50]).

Theorem 26 ([3]). *Let* $f \in S_{1/2}$. *Then for every* $z_0 \in D$,

$$\mathrm{Re}\, \frac{z_0}{f(z_0)}\ \frac{f(z) - f(z_0)}{z - z_0} > 1/2\ ,\quad z \in D\ .$$

Theorem 27 ([50]). *Let* f *be univalent and convex in* D *with* $f(0) = 0$. *Then for every* $z_0 \in D$,

$$z\frac{f(z) - f(z_0)}{z - z_0} \in S_{1/2}\ .$$

Finally, let f be a function starlike in D and let

$$L_r = \int_{-\pi}^{\pi} |\, f'(re^{i\theta})\, |\, r\, d\theta\ ,\quad 0 \leq r < 1\ .$$

Then L_r is the length of the image of the circle $|z| = r$ under the mapping f. In 1959 F.R. Keogh ([51]) showed that if, in addition, f is bounded, then

$$L(r) = O(\log \frac{1}{1-r}) \quad \text{as} \quad r \to 1$$

(see also [52, 53]), and raised the question whether the O here could be reduced to o. It was shown that such a reduction occurs if '$\arg f(e^{i\theta})' = \lim_{r\to 1} \arg f(re^{i\theta})$ is an absolutely continuous function of θ, and that the estimate

$$L(r) = o\,(\log \frac{1}{1-r}) \quad \text{as} \quad r \to 1$$

for that case is best possible in the sense that, corresponding to each function $\eta(r)$ such that $\eta(r) > 0$, $0 < r < 1$, $\lim_{r \to 1} \eta(r) = 0$, a bounded starlike f can be constructed for which $\arg f(e^{i\theta})$ is absolutely continuous and

$$L(r_n) \,/\, \log \frac{1}{1 - r_n} > \eta(r_n)$$

for a sequence $r_n \nearrow 1$.

In 1961 W.K. Hayman ([54]) took up the question raised in [51], and proved the existence of a bounded starlike f for which

$$\limsup_{r \to 1} \left[L(r)/\log \frac{1}{1-r}\right] > 0 .$$

Hayman left open the further question whether the lim sup here can be replaced by lim inf. J.L. Lewis has now shown that indeed it can.

Theorem 28 ([55]). *There exists a bounded starlike* f *such that*

$$\liminf_{r \to 1} \left[L(r)/\log \frac{1}{1-r}\right] > 0 .$$

8. Φ-like Functions. In [56] L. Brickman introduced the class of Φ-like functions. A function f, analytic in D and satisfying the conditions $f(0) = 0$, $f'(0) = 1$, is called Φ-like if

$$\mathrm{Re}\, \frac{z f'(z)}{\Phi(f(z))} > 0 , \quad z \in D ,$$

where $\Phi(w)$ is analytic in $f(D)$, $\Phi(0) = 0$, $\mathrm{Re}\,\Phi'(0) > 0$. Using methods from the theory of differential equations he proved the remarkable result that every Φ-like functions is univalent in D and, conversely, that every function in S is Φ-like for an appropriate Φ. The proof of the converse is elementary. Let $z = \phi(w)$ be the inverse mapping of $w = f(z)$, and let $g(z)$ be *any* analytic function such that $\mathrm{Re}\, g(z) > 0$ in D. Let $h(w) = g(\varphi(w))$, and define

$$\Phi(w) = \frac{\phi(w)}{\phi'(w)h(w)} .$$

It is now easy to verify that f is Φ-like with this definition of Φ.

In [57] St. Ruscheweyh begins by proving

Theorem 29. *Let* $G(z)$ *be a convex conformal mapping of* D, $G(0) = 1$, *and let*

$$F(z) = z \exp \left(\int_0^z \frac{G(x) - 1}{x}\, dx\right).$$

Let $f(z)$ *be analytic in* D, $f(0) = 0$, $f'(0) = 1$. *Then we have*

$$\frac{z f'(z)}{f(z)} \prec G(z) , \quad z \in D ,$$

if and only if for all $|s| \leq 1$, $|t| \leq 1$,

$$\frac{t\, f(sz)}{s\, f(tz)} \prec \frac{t\, F(sz)}{s\, F(tz)}$$

holds.

Here the symbol $\prec$ is as defined in Section 2. This theorem contains earlier results of several authors. Its proof makes use of the Theorem 4 of Ruscheweyh and Sheil-Small stated in Section 2. A special choice of G(z) leads to the following corollary (compare Z. Lewardowski ([58]; p. 123)).

Corollary 1. *Let* f(z) *be analytic in* D, f(0) = 0, f'(0) = 1. *Then we have*

$$\mathrm{Re}\,[e^{i\alpha}\frac{zf'(z)}{f(z)}] > \beta\,,\quad z \in D,$$

for a certain $\alpha \in (-\frac{1}{2}\pi, \frac{1}{2}\pi)$, $\beta < \cos\alpha$, *if and only if*

$$\frac{t\,f(sz)}{s\,f(tz)} \prec (\frac{1-tz}{1-sz})^{2(\cos\alpha-\beta)\exp(-i\alpha)}$$

for all $|s| \leq 1, |t| \leq 1$.

We remark that the condition

$$\mathrm{Re}[e^{i\alpha}\frac{z\,f'(z)}{f(z)}] > 0\,,\ z \in D\,,$$

is definitive of the class of functions f (subject to f(0) = 0, f'(0) = 1) that are *spiral-like* (of type α) in D (see [59]). Such functions are necessarily in S. The case $s = 1, t = \alpha = \beta = 0$ gives the well-known result $f(z)/z \prec (1-z)^{-2}$ of A. Marx ([47]) for starlike $f \in S$.

Ruscheweyh now introduces a definition of Φ-like functions slightly more general than the original one of Brickman. Let G(z) be a convex conformal mapping of D, G(0) = 1. Let f(z) be analytic in D, f(0) = 0, f'(0) = 1. Let $\Phi(w)$ be analytic in f(D), $\Phi(0) = 0$, $\Phi'(0) = 1$, $\Phi(w) \neq 0$ in $f(D)\setminus\{0\}$. Then f(z) is called *Φ-like with respect to* G if

$$\frac{z\,f'(z)}{\Phi(f(z))} \prec G(z),\quad z \in D\,.$$

With this definition and with the aid of Theorem 29, Ruscheweyh next proves

Theorem 30. *Let* G(z), f(z), $\Phi(w)$ *be as in the above definition, and put*

$$Q(w) = w\exp\left(\int_0^w \left(\frac{1}{\Phi(x)} - \frac{1}{x}\right)dx\right)\,,$$

$$F(z) = z\,\exp\left(\int_0^z \frac{G(x)-1}{x}\,dx\right)\,.$$

Then f(z) *is* Φ*-like with respect to* G *if and only if, for all* $|s| \leq 1, |t| \leq 1$,

$$\frac{Q(f(sz))}{Q(f(tz))} \prec \frac{F(sz)}{F(tz)}\,. \tag{13}$$

The complete proof of this theorem is short enough to be given here.

(i) Suppose that f(z) is Φ-like with respect to G, and define

$$H(z) = z\,\exp\left(\int_0^z \left(\frac{f'(x)}{\Phi(f(x))} - \frac{1}{x}\right)dx\right)\,,$$

so that

$$\frac{z\,f'(z)}{\Phi(f(z))} = \frac{zH'(z)}{H(z)} . \tag{14}$$

A simple calculation proves the relation

$$\frac{t\,Q(f(sz))}{s\,Q(f(tz))} = \frac{t\,H(sz)}{s\,H(tz)},\ z \in D. \tag{15}$$

and since $z\,H'(z)/H(z) \prec G(z)$, Theorem 29 gives

$$\frac{t\,H(sz)}{s\,H(tz)} \prec \frac{t\,F(sz)}{s\,F(tz)} ,$$

and the result follows.

(ii) If (13) holds for all $|s| \leq 1, |t| \leq 1$, then by (14), (15) and Theorem 29,

$$z\frac{H'(z)}{H(z)} \prec G(z)$$

and the conclusion follows.

The relation (15) leads to a very short proof of the following result of Brickman ([56]).

Corollary 2. *Let* $f(z)$ *be* Φ*-like with respect to* G , *where* $0 \notin G(D)$. *Then* $f(z)$ *is univalent in* D.

If $0 \notin G(D)$, then we have $z\,f'(z)/\Phi(f(z)) \prec G(z) \prec (1+xz)/(1-z)$ for a certain $x, |x| = 1$. Thus $H(z)$ is spiral-like, and therefore univalent, and from (15) we obtain the univalence of $Q(f(z))$, and consequently of $f(z)$.

Finally, Ruscheweyh proves the following theorem, the proof of which makes use only of classical distortion theorems.

Theorem 31. *If* $f \in S$, *then, for* $0 \leq t \leq 1$,

$$\frac{(1+t)^2}{4t} f(tz) \prec f(z) ,\quad z \in D . \tag{16}$$

Recently P.J. Eenigenburg and J. Waniurski [(60)] have extended this theorem and have given an example of a locally univalent function $f(z)$ of infinite valence satisfying (16).

References

1. G. Pólya and I.J. Schoenberg, Remarks on de la Vallée Poussin means and convex conformal maps of the circle, Pacific J. Math. 8(1958), 295–334.

2. T.J. Suffridge, Convolutions of convex functions, J. Math. Mech. 15(1966), 795–804.

3. S. Ruscheweyh and T. Sheil-Small, Hadamard products of schlicht functions and the Pólya-Schoenberg conjecture, Comment. Math. Helvet.48(1973), 119–135.

4. M.S. Wilf, Subordinating factor sequences for convex maps of the unit circle, Proc. Amer. Math. Soc. 12 (1961), 689–693.

5. L. Bieberbach, Über die Koeffizienten derjenigen Potenzreihen, welche eine schlichte Abbildung des Einheitskreises vermitteln, K. Preuss. Wiss. Berlin, Sitzungsberichte 138(1916), 940–955.

6. K. Loewner, Untersuchungenüber schlichte konforme Abbildungen des Einheitskreises, I. Math. Ann. 89(1923), 103–121.

7. P. Garabedian and M. Schiffer, A proof of the Bieberbach conjecture for the fourth coefficient, J. Rat. Mech. Anal. 4(1955), 427–465.

8. R. Pederson and M. Schiffer, A proof of the Bieberbach conjecture for the fifth coefficient, Arch. Rat. Mech. Anal. 45(1972), 161–193.

9. R. Pederson, A proof of the Bieberbach conjecture for the sixth coefficient, Arch. Rat. Mech. Anal. 31 (1968), 331–351.

10. I. Milin, A bound for the coefficients of schlicht functions, Dokl. Akad. Nauk. SSSR 160(1965), 769–771 (English translation: Soviet Math. Dokl. 6(1965), 196–198).

11. J.E. Littlewood, Lectures on the theory of functions, Oxford University Press, 1944.

12. C.H. FitzGerald, Quadratic inequalities and coefficient estimates for schlicht functions, Arch. Rat. Mech Anal. 46(1972), 356–368.

13. H. Grunsky, Koeffizientenbedingungen für schlicht abbildende meromorphe Funktionen, Math. Zeit. 45(1939),29–61.

14. W.K. Hayman, The asymptotic behaviour of p-valent functions, Proc. London Math. Soc.(3)5(1955), 257–284.

15. D. Horowitz, A refinement for coefficient estimates of univalent functions, Proc. Amer. Math. Soc. 54 (1976), 176–178.

16. G. Goluzin, On distortion theorems and the coefficients of univalent functions, Mat. Sbornik N.S.23 (65)(1948), 353–360.

17. D. Horowitz, Applications of quadratic inequalities in the theory of univalent functions, Doctoral dissertation, University of California, San Diego, 1974.

18. D. Horowitz, Journal d' Analyse. To appear.

19. J. Dieudonné, Sur les fonctions univalentes, C.R. Acad. Sci. Paris 192 (1931), 1148–1150.

20. Z. Nehari, A proof of $|a_4| \leq 4$ by Loewner's method, Symposium on Complex Analysis, Canterbury 1973, London Mathematical Society Lecture Note Series 12, 107–110.

21. Z. Charzynski and M. Schiffer, A new proof of the Bieberbach conjecture for the fourth coefficient, Arch. Rat. Mech. Anal. 5(1960), 187–193.

22. M. Ozawa, On the Bieberbach conjecture for the sixth coefficient, Kodai Math. Sem. Rep. 21(1969), 97–128.

23. A. Baernstein, Proof of Edrei's spread conjecture, Proc. London Math. Soc. (3)26(1973), 418–434.

24. A. Baernstein, Integral means, univalent functions and circular symmetrization, Acta Math. 133(1974), 139–169.

25. I.E. Bazilevič, on distortion theorems and coefficients of univalent functions (Russian), Mat. Sb. N.S., 28, 70(1951), 147–164.

26. Jinfu Feng and T.H. Macgregor, Estimates on integral means of the derivatives of univalent functions, Journal d'Analyse 29(1976), 203–231.

27. O. Lehto, On the distortion of conformal mappings with bounded boundary rotation, Ann. Acad. Sci. Fenn. Ser. A 1 Math. Phys. 124(1952), 14 pp.

28. V. Paatero, Über die konforme Abbildung von Gebieten deren Rander von beschrankter Drehung sind, Ann. Acad. Sci. Fenn. Ser. A9(1931), 77 pp.

29. M. Schiffer and O. Tammi, On the fourth coefficient of univalent functions with bounded boundary rotation, Ann. Acad. Sci. Fenn. Ser. A 396(1967), 26 pp.

30. H. Lonka and O. Tammi, On the use of step-functions in extremal problems of the class with bounded boundary rotations, Ann. Acad. Sci. Fenn. Ser. A 1 418(1968).

31. D.A. Brannan, On functions of bounded boundary rotation II, Bull. London Math. Soc. 1(1969), 321–322.

32. H.B. Coonce, On functions of bounded boundary rotation, Trans. Amer. Math. Soc. 157(1971),39-51.

33. J. Noonan, Asymptotic behaviour of functions with bounded boundary rotation, Trans. Amer. Math. Soc. 164(1972), 397–410.

34. D.A. Brannan, J.G. Clunie and W.E. Kirwan, On the coefficient problems for functions of bounded boundary rotation, Ann. Acad. Sci. Fenn. Ser. A1 Math., (1973), no. 523, 18 pp.

35. D. Aharonov and S. Friedland, On an inequality connected with the coefficient conjecture for functions of bounded boundary rotation, Ann. Acad. Sci. Fenn. A1 524(1972), 13 pp.

36. D.A. Brannan, On coefficient problems for certain power series, Symposium on Complex Analysis, Canterbury 1973, London Math. Soc. Lecture Note Series 12, 17–27.

37. I.M. Milin, Hayman's regularity theorem for the coefficients of univalent functions, Dokl. Acad. Nauk SSSR 192(1970), 738–741 (in Russian).

38. I.M. Milin, Univalent Functions and Orthonormal Systems (Izdat. "Nauka", Moscow, 1971; in Russian).

39. P. Duren, Estimation of coefficients of univalent functions by a Tauberian remainder theorem, J. London Math. Soc. (2), 8(1974), 279–282.

40. I.E. Bazilevič, On a univalence criterion for regular functions and the dispersion of their coefficients, Mat. Sb., 74(1967), 133–146 (in Russian).

41. G.H. Hardy, Divergent Series (Oxford University Press, 1949).

42. I.E. Bazilevič, On the dispersion of the coefficients of univalent functions, Mat. Sb., 68(1965), 549–560 (in Russian).

43. G. Freud, Restglied eines Tauberschen Satzes, I, Acta Math. Acad. Sci. Hungar.,2(1951), 299–308 ; II, ibid.3 (1952), 299–307.

44. J. Korevaar, Best L_1 approximation and the remainder in Littlewood's theorem, Nederl. Akad. Wetensch. Proc. Ser. A.56 = Indagationes Math., 15(1953), 281–293.

45. T.J. Suffridge, Starlike functions as limits of polynomials, Advances in complex function theory, Lecture Notes in Maths., 505, Springer, Berlin-Heidelberg-New York, 1976, 164–202.

46. T.J. Suffridge, Extreme points in a class of polynomials having univalent sequential limits, Trans. Amer. Math. Soc. 163(1972), 225–237.

47. A. Marx, Untersuchungen über schlichte Abbildungen, Math. Ann. 107(1932/33), 40–67.

48. E. Strohhäcker, Beiträge zur Theorie der schlichten Funktionen, Math. Z. 37(1933), 356–380.

49. S. Ruscheweyh, Linear operators between classes of prestarlike functions,(to appear).

50. T.J. Suffridge, Convolutions of convex functions, J. Math. Mech., 15(1966), 795–804.

51. F.R. Keogh, Some theorems on conformal mapping of bounded star-shaped domains, Proc. London Math. Soc., 9(1959),481–491.

52. D.K. Thomas, On starlike and close-to-convex univalent functions, J. London Math. Soc.,42(1967), 427–435.

53. D.K. Thomas, On Bazilevič functions, Trans. Amer. Math. Soc., 132(1968), 353–361.

54. W.K. Hayman, On functions with positive real part, J. London Math. Soc., 36(1961), 35–48.

55. J.L. Lewis, Note on an arc length problem, J. London Math. Soc. (2), 12(1976), 469–474.

56. L. Brickman, Φ-like analytic functions, I, Bull. Amer. Math. Soc., 79(1973), 555–558.

57. St. Ruscheweyh, A subordination theorem for Φ-like functions, J. London Math. Soc.(2),13(1976), 275–280.

58. Z. Lewandowski, Sur certaines classes de fonctions univalentes dans le cercle-unité, Ann. Univ. M. Curie–Skl., Sectio A, 13(1959), 115–126.

59. L. Spacek, Contribution à la théorie des fonctions univalentes , Casopis Pest. Mat. 62(1932),12–19.

60. P.J. Eenigenburg and J. Waniurski, On subordination and majorization, Proc. Amer. Math. Soc., To appear.

ON THE IMAGINARY VALUES OF MEROMORPHIC FUNCTIONS

J. Miles and D. Townsend

For f meromorphic in the plane, let $\varphi(r,f) = \varphi(r)$ be the number of distinct θ in $[0,2\pi)$ such that $\operatorname{Re} f(re^{i\theta}) = 0$. We study the relationship between $\varphi(r)$ and the rate of growth of f. For entire functions it has been shown independently in [2] and [4] that

$$\limsup_{r\to\infty} \frac{\log \varphi(r)}{\log r} = \rho \,, \tag{1}$$

where ρ is the order of f. One advantage of extending the study to meromorphic functions is that the results, applied to the functions $zf'(z)/f(z)$ and $zf''(z)/f'(z) + 1$ respectively, yield information on the number of sign changes in $[0,2\pi)$ of the derivative of $\arg f(re^{i\theta})$ and of the derivative of the argument of the vector tangent to the curve $f(re^{i\theta})$. Such information can lead to results in value distribution theory; see (3).

Theorem. *Let* f *be meromorphic in the plane of order* ρ *and let* $\varphi(r)$ *be as above. Then*

(i) $\varphi(r) < \infty$ *for all but at most one value of* r ;

(ii) *there exists* $E \subset [0,\infty)$ *with* $\int_E x^p\,dx < \infty$ *for all* $p > 0$ *such that*

$$\limsup_{r\to\infty;\, r\notin E} \frac{\log \varphi(r)}{\log r} \leqslant \rho \;;$$

(iii) *if* $\Phi(r) = \int_1^r \frac{\varphi(t)}{t}\,dt$, *then*

$$\limsup_{r\to\infty} \frac{\log \Phi(r)}{\log r} \leqslant \rho \;;$$

and

(iv) *corresponding to any functions* $\psi(r)$ *and* $\alpha(r)$ *satisfying* $\psi(r) \to \infty$ *and* $\alpha(r)/\log r \to \infty$, *there exists a meromorphic* f *with Nevanlinna characteristic* $T(r,f) \leqslant \alpha(r)$ *for all* r *and with*

$$\limsup_{r\to\infty} \frac{\varphi(r)}{\psi(r)} = \infty \,.$$

We remark that, unlike the entire function case, for meromorphic functions we cannot expect lower bounds for $\varphi(r)$ related to the growth of f. This is clear from a consideration of $f(z) = (e^z + 1)/(e^z - 1)$, for which $\varphi(r,f) \equiv 2$.

We indicate the ideas involved in the proof of the theorem.

To establish (i), we simply note that if $\varphi(r_0) = \infty$, then $(f(z) + 1)/(f(z) - 1)$ is a constant of modulus one multiplied by a quotient of properly normalized finite Blaschke products. From this it follows that $\varphi(r) < \infty$ if $r \neq r_0$.

An adaption of the techniques in [2] and [4] giving the upper bound for $\varphi(r)$ in (1) yields (ii) and (iii). From [5, p. 4], if $f(z) = \Sigma_k a_k z^k$ is entire of order ρ, then $f^*(z) = \Sigma_k |a_k| z^k$ is also of order ρ. Let $a_k = \alpha_k + i\beta_k$ with α_k and β_k real. Following [4], observe that the elementary inequality

$$|\alpha_k \cos k\theta - \beta_k \sin k\theta| r^k \leqslant 2|a_k|(e^{\theta_0} r)^k, \qquad |\mathrm{Im}\,\theta| \leqslant \theta_0,$$

implies that the series $\Sigma_k (\alpha_k \cos k\theta - \beta_k \sin k\theta) r^k$ converges uniformly on compact subsets of the complex θ-plane to an entire function u_r satisfying

$$(2) \qquad |u_r(\theta)| \leqslant 2f^*(e^{\theta_0} r), \qquad |\mathrm{Im}\,\theta| \leqslant \theta_0 .$$

The upper bound in (1) on the number of zeros of u_r in $[0,2\pi]$ results from the use of (2) in Jensen's theorem applied to the function u_r on a disc D_r in the complex θ-plane and from the trivial estimate that the number of real zeros of u_r on D_r does not exceed the number of complex zeros. It is important to note that a point θ_r in $[0,2\pi]$ with $|u_r(\theta_r)| \geqslant 1$ may always be chosen as the center of D_r, thus rendering the term $\log |u_r(\theta_r)|$ in Jensen's theorem harmless.

Direct attempts to carry out this type of argument for meromorphic functions meet with certain difficulties. If we denote by h_r the entire functions of θ encountered in such attempts (corresponding to u_r above), we have upper bounds for $|h_r(\theta)|$ analogous to (2) but lack lower bounds for $|h_r(\theta)|$ for any θ in $[0,2\pi]$ and are thus unable to control the term $\log |h_r(\theta_r)|$ in Jensen's theorem. This difficulty can be overcome by the following argument. First we establish that it is sufficient to prove the result for the function $g(z) = zf''(z)/f'(z) + 1$. In fact, if $\varphi(r,f)$ violates either (ii) or (iii), it is a consequence of an estimate [6, p. 386] for the total variation of the argument of the vector tangent to the curve $f(re^{i\theta})$ that $\varphi(r,g)$ also violates either (ii) or (iii). For r with $f'(re^{i\theta}) \neq 0,\infty$ for $0 \leqslant \theta \leqslant 2\pi$, we are guaranteed the existence of θ_r in $[0,2\pi]$ with $|\mathrm{Re}\, g(re^{i\theta_r})| \geqslant \frac{1}{4}$. This fact, in conjunction with a standard minimum modulus theorem [5, p. 21], enables us to control the terms $\log |h_r(\theta_r)|$ sufficiently well to obtain (ii) and (iii).

To construct the desired example for (iv), first note that if $|a_n| = |b_n| < r$ for $1 \leqslant n \leqslant N$ and

$$g_1(z) = \frac{\prod_{n=1}^{N}\left(1 - \frac{z}{a_n}\right)}{\prod_{n=1}^{N}\left(1 - \frac{z}{b_n}\right)},$$

$$g_2(z) = \frac{g_1(z) \prod_{n=1}^{N}\left(1 - \frac{z\bar{b}_n}{r^2}\right)}{\prod_{n=1}^{N}\left(1 - \frac{z\bar{a}_n}{r^2}\right)},$$

and

$$g_3(z) = \frac{g_2(z)\left(1 - \left(\frac{z}{R}\right)^L\right)}{1 + \left(\frac{z}{R}\right)^L},$$

where $R > r$ and L is a positive integer, then $|g_2(re^{i\theta})| \equiv 1$ and the image of $|z| = r$ under g_3 crosses $|w| = 1$ exactly $2L$ times. Realizing that g_3 is again of the form g_1 and that we have freedom to choose R and L as large as we please, we see the construction may be continued to obtain a function f which has arbitrarily slow transcendental growth and yet maps a sequence of circles $|z| = r_n$ onto curves crossing $|w| = 1$ with preassigned frequency. A linear fractional transformation of this f satisfies all conditions of (iv).

An examination of the proof of (ii) and (iii) shows that Problem 2.44 of [1, p. 161] with $\leqslant$ replacing $=$ has an affirmative answer.

A refinement of the upper bound for $\varphi(r, zf'(z)/f(z))$ contained in (ii) yields a second proof of Theorem 1 of [6], which asserts the existence of an absolute constant K such that for any meromorphic f and any $a_1, \ldots, a_q$ in the extended plane,

(3) $$\liminf_{r \to \infty} \sum_{j=1}^{q} |n(r,a_j) - A(r)| < KA(r),$$

where $n(r,a)$ is the number of solutions of $f(z) = a$ counting multiplicity in $|z| \leqslant r$ and $A(r)$ is the average of $n(r,a)$.

Results analogous to (i), (ii), and (iii) for meromorphic functions in the unit disk have been obtained. We do not know whether the exceptional sets in these results are nonempty or even whether a result for holomorphic functions analogous to (1) holds.

References

1. J. Clunie and W. K. Hayman, editors, Symposium on Complex Analysis, *London Mathematical Society Lecture Note Series 12,* 1973.

2. A. Gelfond, Über die harmonischen Funktionen, *Trudy. Fiz.-Matem. Inst. Steklova* **5** (1934), 149–158.

3. S. Hellerstein, The distribution of values of a meromorphic function and a theorem of H. S. Wilf, *Duke Math. Journal* **32** (1965), 749–764.

4. S. Hellerstein and J. Korevaar, The real values of an entire function, *Bull. Amer. Math. Soc.* **70** (1964), 608–610.

5. B. Ja. Levin, Distribution of zeros of entire functions, *A.M.S. Trans.*, Providence, R. I., 1964.

6. J. Miles, Bounds on the ratio $n(r,a)/S(r)$ for meromorphic functions, *Trans. Amer. Math. Soc.* **162** (1971), 383–393.

7. H. S. Wilf, The argument of an entire function, *Bull. Amer. Math. Soc.* **67** (1961), 488–489.

University of Illinois
Urbana, Illinois 61801

THE POINTS OF MAXIMUM MODULUS OF A UNIVALENT FUNCTION

George Piranian

1. The Problem.

For each function f holomorphic in the complex plane or in the unit disk D, let $\nu(r,f)$ $(0 < r < \infty$ or $0 < r < 1)$ denote the number of points on the circle $|z| = r$ where $|f(z)|$ attains its maximum value $M(r,f)$. About 1964, P. Erdös asked whether the counting function $\nu(r,f)$ can be unbounded, also, whether it can tend to infinity, if f is not of the form $f(z) = Az^n$. After the function-theoretic conference of 1964 at Imperial College, W. K. Hayman put the two questions on record [2, Problem 2.16]. In a joint paper [3], F. Herzog and I showed that $\nu(r,f)$ is unbounded in the case of some nontrivial entire functions and in the case of some nontrivial univalent functions in D.

The radial limit $f(e^{i\theta})$ of the univalent function f described in [3] exists at each point $e^{i\theta}$; but the paper [3] makes no provision for the continuity of f on the unit circle C. The present note is a preliminary and tentative report on the construction of a univalent function f, continuous on the closure of D, for which $\nu(r,f) \to \infty$ as $r \to 1$.

Let the domain G in the w-plane consist roughly of a vertical rod of length 1 (bisected by the imaginary axis, and containing in its lower half the origin), of two rods of length ¼ attached at the upper end of the first rod, of four rods of length $1/16$, and so forth (see Figure 1 for a stylized representation). Let the function f map the unit disk D conformally onto G, and let it be normalized so that $f(0) = 0$ and $f'(0) > 0$. Because ∂G is a Jordan curve, f has a continuous extension to the closure of D. Indeed, by a theorem of L. Fejér [1, p. 123], the rectifiability of ∂G implies that the power series of f converges absolutely on C.

For each r in (0,1), let C_r denote the circle $|z| = r$. Figure 1 makes it geometrically obvious that to each positive integer m there corresponds a value r_m $(0 < r_m < 1)$ such that on the circle C_r the modulus $|f(z)|$ has exactly 2^m local maxima. To make them equal, we must modify the domain G. For example, we must change slightly the lengths of some of the rods, so that all the local maxima of $|f(e^{i\theta})|$ are equal. To achieve a slight increase in a local maximum of $|f(re^{i\theta})|$, for a fixed value r less than 1, we can increase slightly the thickness of the corresponding rod, at an appropriate distance from the point $f(0)$.

The management of the details requires an analytic representation of the mapping function f. Instead of constructing a precise mapping function for the domain G, we begin with an analytic expression that represents a reasonable approximation to the mapping function for G. Then we make successive modifications, not to adjust the function to the domain G, but to achieve equality among the local maxima of $|f(z)|$ on more and more of the circles C_r. The discrete representation and construction sketched in Section 2 suggests a continuous method, and we discuss this briefly in Section 3.

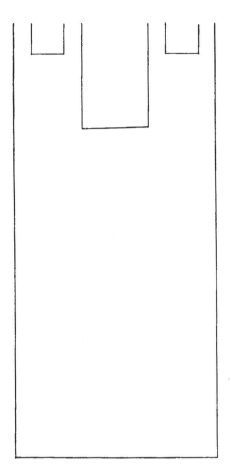

Figure 1.

2. The Discrete Method.

In our first approximation, we use the building block

$$g_n(z) = z_n \frac{\log(1 - z/z_n)}{\log(1 - 1/|z_n|)} ,$$

where z_n denotes a point lying outside of the unit circle C, and where the logarithmic branch is determined by the rule $\log 1 = 0$. We observe that

$$g_n(z) = \frac{z_n}{|\log(1 - 1/|z_n|)|} \left(\frac{z}{z_n} + \frac{1}{2} \left(\frac{z}{z_n} \right)^2 + \cdots \right) ,$$

and that, therefore, for each r the maximum of $|g_n(re^{i\theta})|$ occurs where $e^{i\theta} = z_n/|z_n|$. We note further that

$$g_n\left(\frac{z_n}{|z_n|}\right) = z_n$$

and that on the unit circle C the imaginary part of $z_n^{-1}\, g_n(z)$ is less than

$$\frac{\pi/2}{|\log(1 - 1/|z_n|)|} \quad .$$

It follows that if $|z_n|$ is not much greater than 1, then the value of $f(z)$ lies close to the line through the two points $w = 0$ and $w = z_n$.

Now let

$$\{b_n\} = \left\{1, \frac{1}{4}, \frac{1}{4}, \frac{1}{16}, \frac{1}{16}, \frac{1}{16}, \frac{1}{16}, \frac{1}{64}, \ldots\right\} \quad ,$$

and let

(1) $$f(z) = \Sigma\, b_n g_n(z) \quad .$$

Because the real part of the derivative

$$g_n'(z) = \frac{1}{(1 - z/z_n)\,|\log(1 - 1/|z_n|)|}$$

is positive in D, the univalence of f in the closure of D follows from the theorem of K. Noshiro and S. E. Warschawski (see [4, Theorem 12, p. 151] and [5, Lemma 1, p. 312]). Because $\Sigma\, b_n < \infty$, the series (1) converges uniformly on C, and therefore the function f is continuous on the closure of D. Also, if $z_1 = 2i$, if z_2 and z_3 are symmetric with respect to the imaginary axis and lie close to C, and if the two pairs (z_4, z_5) and (z_6, z_7) are symmetric with respect to the radius vectors of z_2 and z_3, respectively, and so forth, then the image $f(D)$ has the shape of a dichotomously branched tree.

Let r_m denote the value of r for which $M(r, g_n) = \frac{1}{2} M(1, g_n)$, for each index n for which $b_n = 4^{-m}$. By successive approximations, we can modify the sequence $\{b_n\}$ slightly so that on each circle $|z| = r_m$ the equation $|f(z)| = M(r_m, f)$ has 2^m solutions.

Suppose this is accomplished. It remains to choose a dense sequence $\{\rho_j\}$ in $(0,1)$ and to modify the function f further so that $\nu(\rho_j, f)$ is large whenever ρ_j is near 1. Suppose that on the circle $|z| = \rho_1$ the function $|f(z)|$ has 2^m local maxima, and that one of these falls slightly short of $M(\rho_1, f)$. We could increase this local maximum by moving the corresponding point z_n slightly further from C. But it is fairly obvious that no program of shifting the parameters z_n can solve the problem for more than a discrete set of radii ρ_j.

A more promising approach is to split one or more terms of the series (1). Suppose for example that we replace the term $b_n g_n$ with

$$b_{n1} g_{n1} + b_{n2} g_{n2} \quad ,$$

where $b_{n1} + b_{n2} = b_n$ and where the corresponding branch points z_{n1} and z_{n2} lie on the radius vector of z_n. Then the relation

$$b_{n1}g_{n1}(z) + b_{n2}g_{n2}(z) = b_n(z)$$

holds at the point $z = 0$. With appropriate relations between z_1 and z_2, it holds also at the point $z = z_n/|z_n|$. The problem is to develop an algorithm that achieves the necessary adjustments on the circle $|z| = \rho_1$ and creates as little disturbance as possible on the circles $|z| = r_o, r_1, \ldots$. If we can develop an effective algorithm, it will yield a sequence of functions converging to a limit whose counting function has the appropriate values at $r = r_o, r_1, \ldots$ and at $r = \rho_1$. Repeating the process, we find functions whose counting function also has the appropriate values at $r = \rho_2, \rho_3, \ldots$. We can represent the limit function in the form

$$f(z) = \sum_{n=1}^{\infty} \int_1^2 \eta e^{it_n} \frac{\log(1 - z/\eta e^{it_n})}{\log(1 - 1/\eta)} d\mu_n(\eta) , \tag{2}$$

where e^{it_n} lies on the radius vactor of z_n, and where μ_n denotes a mass distribution such that $\int_1^2 d\mu_n(\eta) = b_n$.

3. The Continuous Method.

The transition from (1) to (2) suggests a generalization of the technique in Section 2. Let T denote a dichotomously branched tree (of finite length) in the rectangle S bounded by the lines $\eta = 1$, $\eta = 2$, $\theta = 0$, and $\theta = 2\pi$ (see Figure 2), and for $k = 0,1, \ldots,$ let

$$f_k(z) = \iint_S \eta e^{it} \frac{\log(1 - z/\eta e^{it})}{\log(1 - 1/\eta)} d\mu_k(\eta,t) , \tag{3}$$

where μ_k denotes a measure supported by T. In particular, let μ_o denote arc length on T. I hope to develop an algorithm for successive modifications of the measure $\mu_k(\eta,t)$ such that the sequence $\{f_k\}$ converges to a function f with the property

$$\lim_{r\to 1} \nu(r,f) = \infty .$$

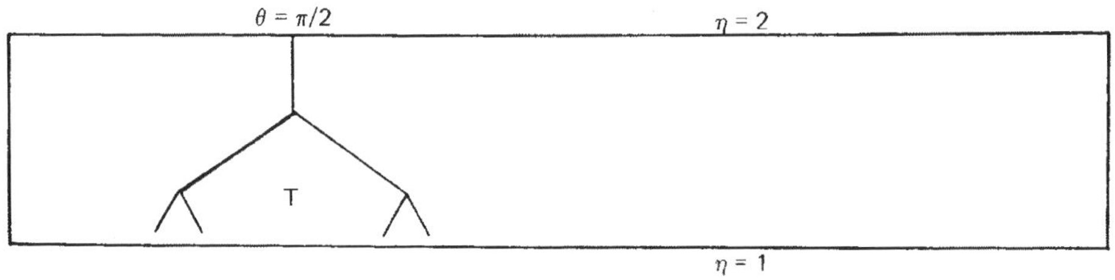

Figure 2.

References

1. L. Fejér, Über gewisse Minimumprobleme der Funktionentheorie, *Math. Ann.* **97** (1927), 104–123.

2. W. K. Hayman, **Research problems in classical function theory**, London Univ. Press, London, 1967.

3. F. Herzog and G. Piranian, The counting function for points of maximum modulus, **Entire Functions and Related Parts of Analysis** (Proc. Symp. Pure Math., La Jolla, Calif., 1966), pp. 240–243, Amer. Math. Soc., Providence, R. I., 1968.

4. K. Noshiro, On the theory of schlicht functions, *J. Fac. Sci.*, Hokkaido Imperial Univ. Sapporo, Ser. I, 2 (1934–1935), 129–155.

5. S. E. Warschawski, On the higher derivatives at the boundary in conformal mapping, *Trans. Amer. Math. Soc.* 38 (1935), 310–340.

The University of Michigan
Ann Arbor, Michigan 48109

VALUES SHARED BY AN ENTIRE FUNCTION AND ITS DERIVATIVE

L. A. Rubel* and Chung-Chun Yang

We say that two meromorphic functions f and g share the value a ($a = \infty$ is allowed) if $f(z) = a$ whenever $g(z) = a$ and also $g(z) = a$ whenever $f(z) = a$, counting multiplicities in both cases. A famous theorem of R. Nevanlinna [H, Theorem 2.6, p. 48] (see also [C]) implies that if two non-constant entire functions f and g on the complex plane share four distinct finite values (ignoring multiplicity) then it follows that $f \equiv g$. And the number *four* cannot be reduced. We consider here the special case $g = f'$, the derivative of f, and prove the following result.

Theorem. *If f is a non-constant entire function in the finite complex plane and if f and f' share two distinct finite values (counting multiplicity), then $f' \equiv f$.*

In other words, a derivative is worth two values. We show at the end of the paper that the number *two* of the theorem cannot be reduced. We do not now know whether there is a corresponding result to our theorem if one ignores multiplicities, or if one considers *meromorphic* instead of *entire* functions.

Proof of Theorem. To fix the ideas, we suppose that f and f' share the values a and b where $a = 1$ and $b = 2$. Other choices of a and b make no real difference, except if a or b is zero, in which case the analysis becomes easier, and is left to the reader. We may write then

1) $$\frac{f' - 1}{f - 1} = e^{k_1}$$

2) $$\frac{f' - 2}{f - 2} = e^{k_2}$$

where k_1 and k_2 are entire functions. We solve 1) and 2) for f to get

3) $$f = \frac{1 + e^{k_1} - 2e^{k_2}}{e^{k_1} - e^{k_2}} .$$

We now differentiate both sides of 3) and substitute in 1) to get

4) $$2e^{2k_1} + e^{2k_2} + (k_2' - k_1' - 3)e^{k_1+k_2} - e^{2k_1+k_2} + e^{k_1+2k_2} + k_1'e^{k_1} - k_2'e^{k_2} = 0 .$$

We shall make repeated use of the following lemma of Hiromi and Ozawa [HO, p. 283].

Lemma. Let $a_0(z), a_1(z), \ldots, a_n(z)$ be meromorphic functions and let $g_1(z), \ldots, g_n(z)$ be entire functions. Further suppose that

* The research of the first author was partially supported by a grant from the National Science Foundation.

$$T(r,a_j) = o\left(\sum_{\nu=1}^{n} m(r,e^{g_\nu})\right), \qquad j = 0,1,\ldots,n$$

holds outside a set of finite logarithmic measure. If an identity

$$\sum_{\nu=1}^{n} a_\nu(z)e^{g_\nu(z)} = a_0(z)$$

holds, then we have an identity

$$\sum_{\nu=1}^{n} c_\nu a_\nu(z)e^{g_\nu(z)} = 0$$

where the constants c_ν, $\nu = 1,\ldots,n$, are not all zero.

Now, in 4), divide by e^{k_2} to get

5) $$2e^{2k_1-k_2} + e^{k_2} + (k_2' - k_1' - 3)e^{k_1} - 2e^{2k_1} + e^{k_1+k_2} + k_1'e^{k_1-k_2} = k_2' .$$

We now apply the lemma to get

6) $$c_1e^{2k_1-k_2} + c_2e^{k_2} + c_3(k_2' - k_1' - 3)e^{k_1} + c_4e^{2k_1} + c_5e^{k_1+k_2} + c_6k_1'e^{k_1-k_2} = 0$$

where $c_1,\ldots,c_6$ are constants not all zero. The hypotheses of the lemma are satisfied because, for example, k_1' is the logarithmic derivative of e^{k_1} and we may use [H, Theorem 2.3, p. 41], which includes R. Nevanlinna's fundamental estimate of the logarithmic derivative. Alternatively, S. Bank has suggested that one could apply Clunie's result [H, Exercise (ii), p. 54] that implies that $T(r,k) = o(T(r,e^k))$. But then it follows that $T(r,k') = o(T(r,e^k))$, outside of a suitably small exceptional set, by [H, Theorem 3.1, p. 55]. At any rate, we divide in 6) by e^{k_1} to get

7) $$c_1e^{k_1-k_2} + c_2e^{k_2-k_1} + c_4e^{k_1} + c_5e^{k_2} + c_6k_1'e^{-k_2} = -c_3(k_2' - k_1' - 3) ,$$

and we may use the lemma once more to get

8) $$d_1e^{k_1-k_2} + d_2e^{k_2-k_1} + d_3e^{k_1} + d_4e^{k_2} + d_5k_1'e^{-k_2} = 0$$

for suitable $d_1,\ldots,d_5$. Multiply by e^{k_2} to get

9) $$d_1e^{k_1} + d_2e^{2k_2-k_1} + d_3e^{k_1+k_2} + d_4e^{2k_2} = -d_5k_1'$$

and apply the lemma yet again to get

10) $$u_1e^{k_1} + u_2e^{2k_1-k_2} + u_3e^{k_1+k_2} + u_4e^{2k_2} = 0$$

where u_1,u_2,u_3,u_4 are constants not all zero. Now by successive applications of the lemma, we reach a contradiction, unless possibly one of the five following conditions holds for some constant C:

$$k_1 = k_2 + C, \quad k_2 = C, \quad k_1 = 2k_2 + C, \quad k_2 = 2k_1 + C, \quad k_1 = C .$$

We now rule out these possibilities unless $f' = f$. First, it is easy to see that $k_1 = C$ (and similarly $k_2 = C$) is inconsistent with 1) and 2) unless $d = e^C = 1$, in which case $f' = f$. For if $(f' - 1)/(f - 1) = d$, then $f = (d - 1)/d + be^{dz}$ for some constant b, and hence $f' = bde^{dz}$. This clearly contradicts 2) unless $d = 1$, for we would have

11) $$\frac{f' - 2}{f - 2} = \frac{bde^{dz} - 2}{be^{dz} - 2} = e^{k_2},$$

which is impossible unless $d = 1$, (remember that f is not constant, so that $b \neq 0$).

Next, we rule out $k_1 = k_2 + C$. We go back to 4) to get

12) $$m_1 e^{2k_1} + m_2 e^{3k_1} = k_1' m_4 e^{k_1} .$$

Apply the lemma again, after dividing by e^{k_1} to get

13) $$n_1 e^{k_1} + n_2 e^{2k_1} = 0 ,$$

which implies that k_1 is a constant, which we have seen implies that $f' = f$.

Finally, we see that $k_1 = 2k_2 + C$ (and similarly, $k_2 = 2k_1 + C$) is impossible, unless $f = f'$. From 9) we would get

14) $$d_2 e^{-C} + (d_1 e^C + d_4) e^{2k_2} + d_3 e^C e^{3k_2} = -2d_5 k_2'$$

and apply the lemma for the last time to get

15) $$\ell_1 + \ell_2 e^{2k_2} + \ell_3 e^{3k_2} = 0$$

where ℓ_1, ℓ_2, ℓ_3 are constants not all zero. In other words, $P(e^{k_2}) = 0$ where P is a cubic polynomial, so e^{k_2} is a constant, which we have already ruled out unless $f = f'$. This completes the proof of the theorem.

Finally, it is easy to see that there exists a nontrivial entire function that does share *one* value with its derivative. For example,

$$f(z) = e^{e^z} \int_0^z e^{-e^t}(1 - e^t)dt$$

satisfies $(f' - 1)/(f - 1) = e^z$ so that f and f' share the value 1. This shows that the number *two* of our theorem is best-possible.

References

[C] H. Cartan, Sur les systèmes de fonctions holomorphes à variétés linéaires lacunaires et leurs applications, *Ann. École Norm. Sup.* **64** (Ser 3:45) (1928), 255–346.

[H] W. K. Hayman, **Meromorphic Functions**, Clarendon Press, Oxford, 1964.

[HO] G. Hiromi and M. Ozawa, On the existence of analytic mappings between two ultrahyperelliptic surfaces, *Kōdai Math. Sem. Report* **17** (1965), 281–306.

University of Illinois at Urbana-Champaign and Naval Research Laboratory, Washington, D.C. 20375

HOLOMORPHIC MAPS OF DISCS INTO F-SPACES

Walter Rudin*

At the Conference on Infinite Dimensional Holomorphy held at the University of Kentucky in 1973, D. J. Patil asked whether there exist holomorphic maps f of the open unit disc U into the open unit ball B of any separable complex Banach space, such that the convex hull of $f(U)$ is dense in B. The following theorem shows that one can actually do much better (or worse, depending on one's attitude toward pathology); in particular, convexity plays no role.

Theorem I. *If Ω is a nonempty connected open subset of a separable complex F-space, then there is a function $f: \overline{U} \to \Omega$ such that*

(i) f *is continuous on* $\overline{U} - \{1\}$,

(ii) f *is holomorphic in* U,

(iii) $f(U)$ *is dense in* Ω.

To avoid misunderstanding, here are some of the relevant definitions.

To say that X is a *complex* F-*space* means that X is a topological vector space with complex scalars whose topology is defined by a *complete translation invariant* metric d: for all $x,y,z \in X$, $d(x+z, y+z) = d(x,y)$

Note that (following Banach [1]) I do not include local convexity among the defining properties of an F-space. The metric can always be so chosen that the balls with center at the origin are *balanced*, i.e., that $d(\alpha x, 0) \leqslant d(x,0)$ if $x \in X$ and $|\alpha| \leqslant 1$. (Theorem 1.24 of [3])

A function $f: U \to X$ is said to be *holomorphic* in U if the limit

$$\lim_{\lambda \to \alpha} (\lambda - \alpha)^{-1} [f(\lambda) - f(\alpha)] = f'(\alpha) \tag{1}$$

exists (in the topology of X) for every $\alpha \in U$. (Of course, this definition can be made with U replaced by any open subset of C.)

The proof of Theorem I uses an X-valued version of the Weierstrass approximation theorem (Theorem II) which is easy and probably known, but for which I have no reference.

Definition. If X is a topological vector space (with real or complex scalars), if $x_i \in X$ for $0 \leqslant i \leqslant n$, and if P maps the scalar field into X according to the formula

$$P(\lambda) = x_0 + \lambda x_1 + \cdots + \lambda^n x_n \tag{2}$$

then P is said to be an *X-polynomial.*

As usual, $[0,1]$ denotes the unit interval in $R \subset C$.

---

* Partially supported by NSF Grant MPS 75-06687.

Theorem II. *If* X *is a topological vector space, if* $f\colon [0,1] \to X$ *is continuous, and if* V *is a neighborhood of* 0 *in* X*, then there exists an X-polynomial* P *such that*

$$(3) \qquad f(t) - P(t) \in V \qquad (0 \leqslant t \leqslant 1)$$

and $P(0) = f(0)$, $P(1) = f(1)$.

Proof. There is a balanced neighborhood W of 0 in X such that $W + W + W \subset V$. (See [3], pp. 10, 11.) Since $[0,1]$ is compact, there is a positive integer m such that $f(s) - f(t) \in W$ whenever $|s - t| \leqslant 1/m$. For $i = 0, \ldots, m$, put $y_i = f(i/m)$, and define

$$(4) \qquad g(t) = (i - mt)y_{i-1} + (mt + 1 - i)y_i$$

if $i - 1 \leqslant mt \leqslant i$.

Then $g(i/m) = y_i = f(i/m)$ for $i = 0, \ldots, m$, and g is a continuous map of $[0,1]$ into the *finite-dimensional* subspace Y of X that is spanned by $\{y_0, \ldots, y_m\}$. The classical Weierstrass theorem therefore gives us a Y-polynomial P such that

$$(5) \qquad g(t) - P(t) \in W \qquad (0 \leqslant t \leqslant 1)$$

and $P(0) = g(0) = f(0)$, $P(1) = g(1) = f(1)$.

Now fix $t \in [0,1]$, and choose i so that $i - 1 \leqslant mt \leqslant i$. Then (4) gives

$$(6) \qquad f(t) - g(t) = [f(t) - y_i] + (i - mt)[y_i - y_{i-1}] \; .$$

Our choice of m shows that each of the vectors in brackets lies in W. Since W is balanced, and $|i - mt| \leqslant 1$, it follows that

$$(7) \qquad f(t) - g(t) \in W + W \qquad (0 \leqslant t \leqslant 1) \, .$$

Now (5) and (7) give the desired conclusion.

Proof of Theorem I. For $n = 1,2,3, \ldots$, put $\beta_n = \exp(i/n)$. Then $\{\beta_n\}$ is a sequence of boundary points of U that converges to 1.

Let Ω be a nonempty connected open set in X, a separable complex F-space. Assume $0 \in \Omega$, without loss of generality. Let $\{x_n\}$, $n = 1,2,3, \ldots$, be a countable dense subset of Ω. Being a connected open subset of a topological vector space, Ω is arcwise connected. (In fact, any two points of Ω can be connected, within Ω, by a piecewise linear path.) Hence, there are continuous functions $g_n\colon [0,1] \to \Omega$ with $g_n(0) = 0$ and $g_n(1) = x_n$. Each $g_n([0,1])$ is compact in Ω, hence has positive distance from Ω^c (the complement of Ω relative to X), and therefore Theorem II furnishes X-polynomials P_n which map $[0,1]$ into Ω, such that $P_n(0) = 0$ and $P_n(1) = x_n$.

The set E_n of all $z \in \mathbf{C}$ such that $P_n(z) \in \Omega$ is open and contains $[0,1]$. Hence E_n contains a closed parallelogram H_n with vertices at 0, 1, and $\frac{1}{2} \pm ih_n$, where $h_n > 0$ is chosen sufficiently small. Then $P_n(H_n)$ is a compact subset of Ω.

We now define $s_0\colon \overline{U} \to \Omega$ by $s_0(\lambda) = 0$, and make the following induction hypothesis (which holds when $n = 1$):

$s_{n-1}: \bar{U} \to \Omega$ is continuous, s_{n-1} is holomorphic in U, and

(8) $$s_{n-1}(\beta_k) = \begin{cases} x_k & \text{if } 1 \leq k \leq n-1, \\ 0 & \text{if } k \geq n. \end{cases}$$

Choose $\delta_n > 0$ so that

(9) $$3\delta_n < \text{dist}(s_{n-1}(\bar{U}), \Omega^c)$$

and

(10) $$3\delta_n < \text{dist}(P_n(H_n), \Omega^c).$$

Since $s_{n-1}(\beta_n) = 0$, there is a circular open disc D_n, centered at β_n, with radius $< 1/n^2$, such that

(11) $$d(s_{n-1}(\lambda), 0) < \delta_n \qquad (\lambda \in \bar{D}_n \cap \bar{U}).$$

Since $\{1\} \cup \{\beta_k\}$ is a compact subset of the unit circle, of Lebesgue measure 0, it is a peak-interpolation set for the disc algebra. Thus, there exists $\varphi_n: \bar{U} \to \bar{U}$, continuous, holomorphic in U, such that $\varphi_n(\beta_n) = 1$, $\varphi_n(\beta_k) = -1$ for all $k \neq n$, $\varphi_n(1) = -1$, and $|\varphi_n(\lambda)| < 1$ for all other $\lambda \in \bar{U}$.

(Actually, the proof of Lemma 2 in [2] suffices for the particular case of the peak-interpolation theorem that is needed here.)

For $j = 1,2,3,\ldots$, let $\psi_{n,j}$ be the conformal map of U onto the interior of H_n that sends 1 to 1, -1 to 0, and 0 to $1/j$.

Consider the functions

(12) $$f_{n,j} = P_n \circ \psi_{n,j} \circ \varphi_n$$

that map $\bar{U}$ into $P_n(H_n) \subset \Omega$.

As $j \to \infty$, $\psi_{n,j} \circ \varphi_n \to 0$ uniformly in the complement of any neighborhood of β_n. Hence $f_{n,j}$ maps $\bar{U} - D_n$ into the ball of radius δ_n and center 0 if j is large enough. (Recall that $P_n(0) = 0$.) Also, since the range of P_n lies in a finite-dimensional vector space, the scalar theory shows that $f'_{n,j} \to 0$ uniformly on compact subsets of U, as $j \to \infty$, and that the same is true for the difference quotients. Thus, if j is sufficiently large, and if $f_n = f_{n,j}$, we have the following properties:

f_n is a continuous map of $\bar{U}$ into $P_n(H_n) \subset \Omega$, f_n is holomorphic in U,

(13) $$f_n(\beta_k) = \begin{cases} x_n & \text{if } k = n, \\ 0 & \text{if } k \neq n, \end{cases}$$

(14) $$d(f_n(\lambda), 0) < \delta_n \qquad (\lambda \in \bar{U} - D_n)$$

and

(15) $$d\left(\frac{f_n(\lambda) - f_n(\alpha)}{\lambda - \alpha}, 0\right) < \frac{1}{n^2}$$

if $|\lambda| \leq 1 - 1/n$, $|\alpha| \leq 1 - 1/n$, $\lambda \neq \alpha$.

We now put

(16) $$s_n = s_{n-1} + f_n .$$

Then s_n is continuous on $\bar{U}$, holomorphic in U, (8) holds with $n + 1$ in place of n, and we claim that $s_n(\bar{U}) \subset \Omega$. Indeed, if $\lambda \in \bar{D}_n \cap \bar{U}$, it follows from (10) and (11) that

(17) $$\text{dist}\,(s_n(\lambda), \Omega^c) > 2\delta_n ,$$

since $f_n(\lambda) \in P_n(H_n)$; if $\lambda \in \bar{U} - D_n$, then (17) follows from (9) and (14).

Our induction hypothesis holds therefore with $n + 1$ in place of n. We now choose $\delta_{n+1} > 0$ so that (9) and (10) hold with $n + 1$ in place of n, and so that $2\delta_{n+1} < \delta_n$. The preceding construction can then be repeated.

Finally, we claim that

(18) $$f = \sum_{n=1}^{\infty} f_n$$

has the desired properties.

Fix n, take $\lambda \in \bar{U} - (D_n \cup D_{n+1} \cup D_{n+2} \cup \cdots)$. By (14),

(19) $$\sum_{j=n}^{\infty} d(f_j(\lambda), 0) < \sum_{j=n}^{\infty} \delta_j < 2\delta_n .$$

Hence (18) converges uniformly on $\bar{U} - D$, where D is any neighborhood of 1. Thus f is continuous on $\bar{U} - \{1\}$. Since $f_n(1) = 0$ for all n, $f(1) = 0$.

Since $s_{n-1} = f_1 + \cdots + f_{n-1}$, (9) and (19) imply that

(20) $$\text{dist}\,(f(\lambda), \Omega^c) > \delta_n$$

if $\lambda \in \bar{U} - (D_n \cup D_{n+1} \cup \cdots)$. Thus $f(\bar{U}) \subset \Omega$.

Since $f(\beta_k) = x_k$, by (13), and $\{x_k\}$ is dense in Ω, we see that $f(U)$ is dense in Ω.

To show that f is holomorphic in U, fix $\alpha \in U$, choose n so that $|\alpha| < 1 - 1/n$, then restrict λ so that $|\lambda| < 1 - 1/n$. It follows from (15) that $d(f'_n(\alpha), 0) \leq n^{-2}$, so that all series in the following decomposition converge:

$$\frac{f(\lambda) - f(\alpha)}{\lambda - \alpha} - \sum_{k=1}^{\infty} f'_k(\alpha) = \left\{\frac{s_n(\lambda) - s_n(\alpha)}{\lambda - \alpha} - s'_n(\alpha)\right\} + \sum_{k=n+1}^{\infty} \frac{f_k(\lambda) - f_k(\alpha)}{\lambda - \alpha} - \sum_{k=n+1}^{\infty} f'_k(\alpha) .$$

The sum of each of the last two series lies in the ball of radius $1/n$, center 0. As $\lambda \to \alpha$, the difference in braces tends to 0. It follows that the limit in (1) exists; in fact, we have $f'(\alpha) = \Sigma f'_k(\alpha)$.

This completes the proof.

References

1. S. Banach, Théorie des Opérations linéaires, Warsaw, 1932.

2. W. Rudin, Boundary values of continuous analytic functions, *Proc. Amer. Math. Soc.* **7** (1956), 808–811.

3. W. Rudin, **Functional Analysis**, McGraw-Hill, 1973.

Postscript. After completion of the present paper, I received two preprints from J. Globevnik, in which he proves Theorem I, but for Banach spaces, rather than F-spaces. These papers are entitled "Analytic functions whose range is dense in a ball" and "Analytic extensions of vector-valued functions." They are to appear in J. Functional Anal. and in Pacific J. Math., respectively. Globevnik's proof relies on vector-valued interpolation theorems.

University of Wisconsin
Madison, Wisconsin 53706

ON THE ZEROS OF THE SUCCESSIVE DERIVATIVES OF INTEGRAL FUNCTIONS II

I. J. Schoenberg

Abstract. In 1936 the author proved [2] the following:

Theorem 1. *If* $f(z)$ $(\not\equiv 0)$ *is entire of exponential type* δ *such that each* $f^{(\nu)}(x)$ $(\nu = 0,1,\dots)$ *vanishes somewhere in the interval* $I_1 = [0,½]$ *of the real axis, then*

(1) $$\delta \geqslant \pi ,$$

and the function

(2) $$f(z) = \cos \pi z$$

shows that π *is the best constant, because* $\cos \pi z$ *satisfies all conditions.*

The present note discusses the following:

Conjecture 1. *If* $f(z)$ $(\not\equiv 0)$ *is entire of exponential type* δ *such that* $f^{(\nu)}(x)$ $(\nu = 0,1,\dots)$ *vanishes at least* k *times in the interval* $I_k = [0, k-½]$, *then the inequality* (1) *holds, and the function* $\cos \pi z$, *which satisfies all conditions, shows that* π *is in* (1) *the best constant.*

In [3] I gave a new proof of Theorem 1 based on a property of the Euler spline $\mathcal{E}_n(x)$ in the interval $[0,½]$. A possible proof of Conjecture 1 is here shown to follow from a Conjecture 2 concerning the Euler spline $\mathcal{E}_n(x)$ in the interval I_k.

1. Introduction. In 1935 J. M. Whittaker [5] formulated the problem of finding the least exponential type W of an entire function $f(z)$, $f(z) \not\equiv 0$, such that each derivative $f^{(\nu)}(z)$ $(\nu \geqslant 0)$ has at least one zero in the unit circle $|z| \leqslant 1$. Observing that the function $f(z) = \sin \pi(z+1)/4$ satisfies these conditions, because $f^{(\nu)}((-1)^{\nu+1}) = 0$, $(\nu = 0,1,\dots)$, Whittaker asked whether $\pi/4$ might be the value of the required least type W. In 1936 I showed in [2] that $\pi/4$ is the least type if we restrict the problem by requiring that each $f^{(\nu)}(x)$ vanish at some point of the interval $[-1,1]$ of the axis of reals, i.e. the interval $[-1,1]$ is to replace the unit circle in Whittaker's problem. For more recent work and references on Whittaker's constant W, see Buckholtz's paper [1].

On replacing the interval $[-1,1]$ by $[0,½]$, the main result of [2] may be stated as follows.

Theorem 1. *If* $f(z)$ *is entire of exponential type* δ *such that*

(1.1) $$\text{each } f^{(\nu)}(x) \ (\nu = 0,1,\dots) \text{ vanishes somewhere in } [0,½],$$

then

(1.2) $$\delta \geqslant \pi ,$$

and the function

(1.3) $$f(z) = \cos \pi z$$

shows that π *is the best constant in* (1.2), *because the function* (1.3) *satisfies the conditions* (1.1).

Let k be a natural number. We observe that the function (1.3) has the property that each $f^{(\nu)}(x)$ vanishes in precisely k points of the interval $0 \leqslant x \leqslant k - \frac{1}{2}$. This suggests the following

Conjecture 1. *If* $f(z)$ $(\not\equiv 0)$ *is entire of exponential type* δ *and is such that*

(1.4) *each* $f^{(\nu)}(x)$ $(\nu = 0,1,\ldots)$ *has at least* k *zeros in* $I_k = [0, k-\frac{1}{2}]$,

then (1.2) *holds, and the function* (1.3) *would show that* π *is the best constant in* (1.2).

In Section 3 we define a certain numerical constant $K_{n,k}$ (Definition 2 below), where n and k are natural numbers. In the note [3] I determined the value of the constant $K_{n,1}$ and derived from its value a new proof of Theorem 1. Here we give an upper bound for $K_{n,k}$ (Theorem 2 below), and conjecture that the upper bound of Theorem 2 gives the value of $K_{n,k}$ (Conjecture 2 below). Recently, Dr. Allan Pinkus showed that $K_{n,k} = n!$ if $n \leqslant k$ (Theorem 3 below), and this establishes the truth of Conjecture 2 if $n \leqslant k$ (Corollary 2 below). Finally, it is shown in Theorem 4 that the truth of Conjecture 2 implies the truth of Conjecture 1. The so-called Euler spline $\mathcal{E}_n(x)$, of degree n, plays a major role, and its relevant properties are recalled in Section 2, with a reference to [4].

2. The Euler splines. We need the stylized versions of the function $\cos \pi x$ known as the *Euler splines* $\mathcal{E}_n(x)$. These are piecewise polynomial functions whose importance was demonstrated in the 1930's by Favard and Kolmogorov. For an independent elementary derivation and some references, see [4]. The function $\mathcal{E}_n(x)$ has the following properties:

(2.1) $\mathcal{E}_n(x+1) = -\mathcal{E}_n(x)$, hence also $\mathcal{E}_n(x+2) = \mathcal{E}_n(x)$ $(x \in \mathbf{R})$, so that $\mathcal{E}_n(x)$ is periodic of period 2.

(2.2) $\mathcal{E}_n(x)$ is a polynomial of degree n in each unit interval $(\nu, \nu+1)$ if n is odd, and has this property in each interval $(\nu - \frac{1}{2}, \nu + \frac{1}{2})$ if n is even, for each integer ν.

(2.3) $$\mathcal{E}_n(x) \in C^{n-1}(\mathbf{R}) .$$

(2.4) $$\mathcal{E}_n(\nu) = (-1)^\nu \quad \text{for all integers } \nu .$$

In terms of the supremum norm on $\mathbf{R}$ we have

(2.5) $$\|\mathcal{E}_n\| = 1 .$$

It is known that the properties (2.2), (2.3), (2.4) and (2.5) characterize uniquely the function $\mathcal{E}_n(x)$. The function $\mathcal{E}_n(x)$ is n times differentiable and the graph of $\mathcal{E}_n^{(n)}(x)$ is a so-called "square wave," i.e., $|\mathcal{E}_n^{(n)}(x)|$ is constant in absolute value for all real x different from the "knots" ν, or $\nu + \frac{1}{2}$. The value of

$$\|E_n^{(n)}\| = \gamma_n \tag{2.6}$$

is a rational number which can be expressed in terms of Euler or Bernoulli numbers by

$$\gamma_n = \begin{cases} \dfrac{2^n n!}{|E_n|} & \text{if } n \text{ is even} \\[2ex] \dfrac{(n+1)!}{2(2^{n+1}-1)|B_{n+1}|} & \text{if } n \text{ is odd.} \end{cases} \tag{2.7}$$

Thus, (see [4, 130])

$$\gamma_1 = 2, \quad \gamma_2 = 8, \qquad \gamma_3 = 24, \qquad \gamma_4 = \frac{384}{5} \quad . \tag{2.8}$$

We also mention the Fourier series expansion

$$E_n(x) = \frac{\sum_0^\infty \dfrac{(-1)^{\nu(n+1)}}{(2\nu+1)^{n+1}} \cos(\pi(2\nu+1)x)}{\sum_0^\infty \dfrac{(-1)^{\nu(n+1)}}{(2\nu+1)^{n+1}}} \quad . \tag{2.9}$$

By differentiations this gives the relation

$$\frac{\pi^n}{\gamma_n} = \frac{4}{\pi} \sum_0^\infty \frac{(-1)^{\nu(n+1)}}{(2\nu+1)^{n+1}} \quad . \tag{2.10}$$

From this we obtain the following facts to be used below:

$$\gamma_n < \pi^n \qquad (n = 1,2,\dots), \tag{2.11}$$

$$\lim_{n\to\infty} (\gamma_n)^{1/n} = \pi \; . \tag{2.12}$$

3. A norm inequality. We need a few definitions. Let n and k be natural numbers and let $\ell > 0$. Let $W_\infty^{(n)}[0,\ell]$ denote the space of functions $f(x)$, $(0 \leqslant x \leqslant \ell)$ such that $f(x) \in C^{n-1}[0,\ell]$, while $f^{(n-1)}(x)$ satisfies in $[0,\ell]$ a Lipschitz condition. It follows that $\|f^{(n)}\| = \text{ess. sup } |f^{(n)}(x)|$ is finite. Let $f(x) \in W_\infty^{(n)}[0,\ell]$. Concerning $f^{(\nu)}(x)$, $(\nu \leqslant n-1)$ we say that ξ $(0 \leqslant \xi \leqslant \ell)$ is a zero of $f^{(\nu)}(x)$ of multiplicity $\geqslant r$, provided that we can write $f^{(\nu)}(x) = (x-\xi)^r Q(x)$, $(0 \leqslant x \leqslant \ell)$, where $Q(x) = O(1)$ as $x \to \xi$.

Definition 1. *We define the class*

$F_{n,k}(\ell) = \{f(x)\colon f(x) \in W_\infty^{(n)}[0,\ell],\ \|f\| > 0$ *and such that each* $f^{(\nu)}(x)$, $(\nu = 0,\dots,n-1)$ *has in* $[0,\ell]$ *a set of zeros of total multiplicity* $\geqslant k\}$.

Observe that if $k > 1$, then $f(x) \equiv x^n \notin F_{n,k}(\ell)$, because $f^{(n-1)}(x)$ has in $[0,\ell]$ only one simple zero at $x = 0$, instead of at least k zeros. However, writing $x_+ = \max(x,0)$, observe that

$$f_\epsilon(x) \equiv (x-\epsilon)_+^n \in F_{n,k}(\ell) , \qquad (0 < \epsilon < 1) \; . \tag{3.1}$$

Indeed, any $\xi \in [0,\epsilon)$ may serve as a zero of $f_\epsilon^{(\nu)}(x)$, $(\nu \leqslant n-1)$ of multiplicity $\geqslant k$.

Definition 2. *We define the constant* $K_{n,k}$ *by requiring that*

$$\|f^{(n)}\| \geqslant K_{n,k}\|f\| \qquad \text{for every } f(x) \in F_{n,k}(1), \tag{3.2}$$

and that $K_{n,k}$ *should be the largest constant satisfying* (3.2).

Equivalently, we may define

$$K_{n,k} = \inf_f \frac{\|f^{(n)}\|}{\|f\|} \qquad \text{if } f(x) \in F_{n,k}(1),\ \|f\| > 0. \tag{3.3}$$

Observe that if $f(x) \in F_{n,k}(\ell)$, then $f(\ell x) \in F_{n,k}(1)$, and now (3.2) shows that $\ell^n\|f^{(n)}\| \geqslant K_{n,k}\|f\|$. Therefore

$$\|f^{(n)}\| \geqslant \frac{K_{n,k}}{\ell^n}\|f\| \qquad \text{if } f(x) \in F_{n,k}(\ell), \tag{3.4}$$

and $K_{n,k}$ *is here the best constant.*

If we select in (3.4) the special values $\ell = ½$ and $k = 1$ we obtain that

$$\|f^{(n)}\| \geqslant 2^n K_{n,1}\|f\| \qquad \text{if } f(x) \in F_{n,1}(½). \tag{3.5}$$

The main result of the note [3] *is its Theorem 3* [3, 365], *which is equivalent to the statement that* (3.5) *holds, and that it holds with the equality sign if and only if* $f(x)$ *is of the form*

$$f(x) = CE_n(x) \quad \text{or} \quad f(x) = CE_n(½ - x) \quad \text{in } 0 \leqslant x \leqslant ½. \tag{3.6}$$

This allows $K_{n,1}$ to be expressed in terms of γ_n: From (2.5), (2.6) we obtain in $[0,½]$ that, if $f(x) = E_n(x)$, then $\|f\| = 1$, $\|f^{(n)}\| = \gamma_n$, and (3.5) with the equality sign, shows that $\gamma_n = 2^n K_{n,1}$; hence

$$K_{n,1} = \frac{1}{2^n}\gamma_n. \tag{3.7}$$

The equivalence of our last italicized statement with our old Theorem 3 of [3] is due to the fact that $E_n(x)$, due to its peculiar properties, may also be written explicitly as

$$E_n(x) = \frac{\int_{½}^{x} dx_1 \int_0^{x_1} dx_2 \int_{½}^{x_2} dx_3 \cdots \int_{\alpha}^{x_{n-1}} dx_n}{\int_{½}^{0} dx_1 \int_0^{x_1} dx_2 \cdots \int_{\alpha}^{x_{n-1}} dx_n} \tag{3.8}$$

where the lower limits of integration form the periodic sequence ½, 0, ½, 0, . . . , the last, α say, being 0 or ½ depending on the parity of n. It was shown in [3] that the function (3.8) (we worked there with the interval $[-1,1]$ rather than $[0,½]$) was the function which produced equality in (3.5).

Here is a consequence of the relation (3.7):

Corollary 1. *The constants* $K_{n,k}$ *defined by* (3.3) *are all positive; in fact,*

(3.9) $$K_{n,k} \geqslant K_{n,1} = \frac{1}{2^n}\gamma_n \quad .$$

Indeed, observe that the class $F_{n,k}(1)$ shrinks as k increases, and $K_{n,k}$ being defined by (3.3) as an infimum, can only increase with k.

Let us briefly recall how (3.7) was used in [3] to establish Theorem 1 of our introduction. We know that (3.7) is equivalent to the statement that

(3.10) $$\|f^{(n)}\| \geqslant \gamma_n \|f\| \qquad \text{if } f(x) \in F_{n,1}(\tfrac{1}{2}) \ .$$

On the other hand we know that if $f(z)$ is entire, then its exponential type δ is given by

(3.11) $$\overline{\lim_{n\to\infty}} |f^{(n)}(0)|^{1/n} = \delta \ .$$

From this we easily derive (see [3, Lemma 3, p. 371]) that in $[0,\tfrac{1}{2}]$ we also have

(3.12) $$\overline{\lim_{n\to\infty}} \|f^{(n)}\|^{1/n} = \delta \ .$$

If the entire function $f(z)$ satisfies (1.1) then $f(x) \in F_{n,1}(\tfrac{1}{2})$ for all n, and (3.10) shows that $\|f^{(n)}\|^{1/n} \geqslant \gamma_n^{1/n}\|f\|^{1/n}$, for all n. Using (2.12) and $\|f\|^{1/n} \to 1$, we obtain the conclusion (1.2).

4. Upper bounds for the constant $K_{n,k}$. Having determined $K_{n,1}$ by (3.7), we shall assume that

(4.1) $$k > 1 \ .$$

We have already observed, in formulating our Conjecture 1, that $f(x) = \cos \pi x$ belongs to the class $F_{n,k}(k - \tfrac{1}{2})$ for all values of n. For this class we find from (3.4), for $\ell = k - \tfrac{1}{2}$, that

(4.2) $$K_{n,k} = \frac{\inf\limits_{f} (k - \tfrac{1}{2})^n \|f^{(n)}\|}{\|f\|} \qquad \text{if } f(x) \in F_{n,k}(k - \tfrac{1}{2}) \ .$$

Observe, however, that we obtain a better, i.e. smaller, bound for $K_{n,k}$ if we replace the function $\cos \pi x$ by

(4.3) $$f(x) = \cos \pi x + (-1)^k \in F_{n,k}(k - \tfrac{1}{2}) \ ,$$

which belongs to the same class, and gives $\|f\| = 2$, $\|f^{(n)}\| = \pi^n$. From (4.2) we conclude that

(4.4) $$K_{n,k} \leqslant \tfrac{1}{2}(k - \tfrac{1}{2})^n \pi^n \ .$$

However, also the function

(4.5) $$f(x) = E_n(x) + (-1)^k \in F_{n,k}(k - \tfrac{1}{2})$$

for precisely the same reasons that led to (4.3). Now $\|f\| = 2$, while $\|f^{(n)}\| = \gamma_n$, and (4.2) shows that

(4.6) $$K_{n,k} \leqslant \tfrac{1}{2}(k - \tfrac{1}{2})^n \gamma_n \ .$$

In view of the inequality (2.11) we see that the upper bound (4.6) is better than the bound given by (4.4).

A second upper bound for $K_{n,k}$ is given by the inequality

(4.7) $$K_{n,k} \leqslant n!$$

Indeed, for the function

$$f_\epsilon(x) = (x - \epsilon)_+^n, \qquad (0 \leqslant x \leqslant 1),$$

we find that $f_\epsilon(x) \in F_{n,k}(1)$, while, for its sup norms in $[0,1]$, we have $\|f_\epsilon^{(n)}\| = n!$ and $\|f_\epsilon\| = (1-\epsilon)^n$. From the definition (3.3) we conclude that

$$K_{n,k} \leqslant \frac{n!}{(1-\epsilon)^n} \to n! \qquad \text{as } \epsilon \to 0+ .$$

This establishes (4.7). I owe to Dr. Allan Pinkus the use of the function $f_\epsilon(x)$ for proving (4.7); my own proof was slightly more complicated.

We may combine the inequalities (4.6) and (4.7) in

Theorem 2. *If* $k > 1$, *then the constant* $K_{n,k}$ *of Definition 2 satisfies the inequality*

(4.8) $$K_{n,k} \leqslant \min\{\tfrac{1}{2}(k - \tfrac{1}{2})^n \gamma_n, n!\}.$$

Which of the two bounds (4.6) and (4.7) is the smaller one depends on the relative sizes of n and k. From $\gamma_n < \pi^n$ it is clear that the first is the smaller one if k is fixed and n becomes large, and the opposite holds if n is fixed and we let k become large.

With all due caution I propose as a "working hypothesis" the following

Conjecture 2. *If* $k > 1$ *we have the equality sign in* (4.8); *hence*

(4.9) $$K_{n,k} = \min\{\tfrac{1}{2}(k - \tfrac{1}{2})^n \gamma_n, n!\}.$$

5. The case when $k \geqslant n$. This case was recently settled by Dr. A. Pinkus who established

Theorem 3 (A. Pinkus). *If*

(5.1) $$k \geqslant n$$

then

(5.2) $$K_{n,k} = n!.$$

Proof. Since $K_{n,k}$ is a non-decreasing function of k, we conclude by (4.7) and (5.1)

(5.3) $$n! \geqslant K_{n,k} \geqslant K_{n,n}$$

so that

(5.4) $$K_{n,n} \leqslant n!.$$

Let us establish the opposite inequality

(5.5) $$K_{n,n} \geqslant n!.$$

We consider an arbitrary but fixed

(5.6) $$f(x) \in F_{n,n}(1)\,,$$

and let ξ be such that

(5.7) $$|f(\xi)| = \|f\|\ .$$

From Definition 1 we know that $f(x)$ has at least n zeros in $[0,1]$, and let n of these zeros be x_ν, $x_1 \leqslant x_2 \leqslant \cdots \leqslant x_n$. Let $p_f(x)$ be the polynomial of degree n such that

(5.8) $$p_f(x_\nu) = f(x_\nu) \quad (=0) \quad (\nu = 1, \ldots, n)\,, \qquad p_f(\xi) = f(\xi) \quad (= \pm\|f\|)\,.$$

The last relation (5.8) clearly implies that

(5.9) $$\|p_f\| \geqslant \|f\|\ .$$

From (5.8) we conclude that $f(x) - p_f(x)$ has $n+1$ zeros. Rolle's theorem implies that $f^{(n)}(x) - p_f^{(n)}$ has at least one zero or else a change of sign in $[0,1]$. Since $p_f^{(n)}(x)$ is constant, we conclude that

(5.10) $$\|p_f^{(n)}\| \leqslant \|f^{(n)}\|\ .$$

Finally, (5.9) and (5.10) show that

(5.11) $$\frac{\|f^{(n)}\|}{\|f\|} \geqslant \frac{\|p_f^{(n)}\|}{\|p_f\|}\ .$$

Let $P_n = \{p(x)\}$ denote the class of polynomials $p(x) = x^n$ + lower degree terms, having n zeros in $[0,1]$. Let $p_f(x) = Cp(x)$, where $p(x) \in P_n$. Since $\|p_f^{(n)}\|/\|p_f\| = \|p^{(n)}\|/\|p\|$, we conclude from (5.11) that

(5.12) $$K_{n,n} = \inf_{f \in F_{n,k}(1)} \frac{\|f^{(n)}\|}{\|f\|} \geqslant \inf_{p \in P_n} \frac{\|p^{(n)}\|}{\|p\|}$$

However, *the second infimum in* (5.12) *is attained for* $p = x^n$ *and has therefore the value* $n!$.

Indeed, if $p(x) \in P_n$, hence $p(x) = \Pi_1^n (x - \alpha_\nu)$, $(0 \leqslant \alpha_\nu \leqslant 1$ for all $\nu)$, then $\|p^{(n)}\| = n!$, while $|p(x)| \leqslant 1$. Therefore,

$$\sup_{p \in P_n} \|p\| = 1\,,$$

where this supremum is reached if $p(x) = x^n$.

This establishes (5.5). Finally, (5.4) shows that (5.2) holds.

Corollary 2. *We have the inequality*

(5.13) $$\tfrac{1}{2}(k - \tfrac{1}{2})^n \gamma_n \geqslant n! \qquad \textit{if } k > 1 \textit{ and } k \geqslant n\ .$$

Proof. The opposite inequality would contradict (5.2) in view of (4.6).

Corollary 3. *Conjecture 2 is correct if* $k \geqslant n$.

Proof. By (5.13) and (5.2) we have that

$$\min\{\tfrac{1}{2}(k - \tfrac{1}{2})^n \gamma_n, n!\} = n! = K_{n,k}\ .$$

6. The connection between the two conjectures. We wish to establish our last

Theorem 4. *The truth of Conjecture 2 implies the truth of Conjecture 1.*

Proof. The proof is much like the derivation (sketched at the end of Section 3) of Theorem 1 from (3.7). Let us assume that (4.9) holds and that the entire function $f(z)$ of exponential type δ satisfies the conditions (1.4). It follows that $f(x) \in F_{n,k}(k - \frac{1}{2})$ for every n, and now (3.4) implies that in the interval $[0, k - \frac{1}{2}]$ we have

$$\|f^{(n)}\| \geqslant K_{n,k}(k - \tfrac{1}{2})^{-n}\|f\| \qquad \text{for all } n. \tag{6.1}$$

On the other hand we conclude from (4.9) that *for sufficiently large* n and $k > 1$ we have

$$K_{n,k} = \tfrac{1}{2}(k - \tfrac{1}{2})^n \gamma_n .$$

Now, (6.1) shows that

$$\|f^{(n)}\| \geqslant \tfrac{1}{2}\gamma_n\|f\| \qquad \text{for all sufficiently large } n.$$

Taking the nth roots of both sides and using (3.12) and (2.12) we obtain the desired inequality (1.2).

References

1. J. D. Buckholtz, The Whittaker constant and successive derivatives of entire functions, *J. Approximation Theory* 3 (1970), 194–212.

2. I. J. Schoenberg, On the zeros of successive derivatives of integral functions, *Trans. Amer. Math. Soc.* **40** (1936), 12–23.

3. ____________, Norm inequalities for a certain class of C^∞ functions, *Israel J. of Math.* **10** (1971), 364–372.

4. ____________, The elementary cases of Landau's problem of inequalities between derivatives, *Amer. Math. Monthly* **80** (1973), 121–158.

5. J. M. Whittaker, Interpolatory function theory, *Cambridge Tracts in Math. and Math. Phys.* No. 33, Cambridge, 1935.

Mathematics Research Center
University of Wisconsin–Madison

and

University of Pittsburgh

ENTIRE FUNCTIONS OF BOUNDED INDEX

S. M. Shah

1. Introduction. In this paper we give a survey of known results on entire functions of bounded index and functions of bounded value distribution. Related results on locally multivalent functions and proofs of a few theorems are included. A reference is given for each theorem stated without proof.

In what follows we shall be concerned with entire transcendental functions of one variable, except when indicated otherwise.

Consider the Taylor expansion of an entire function f about a point ω:

$$f(z) = \sum_{n=0}^{\infty} \frac{f^{(n)}(\omega)}{n!}(z-\omega)^n, \qquad \omega \in \mathbf{C}. \tag{1.1}$$

Since $\lim_{n\to\infty} f^{(n)}(\omega)/n! = 0$, there exists a least nonnegative integer N_ω such that

$$\sup_{0\leqslant n<\infty}\left\{\left|\frac{f^{(n)}(\omega)}{n!}\right|\right\} = \left|\frac{f^{(N_\omega)}(\omega)}{(N_\omega)!}\right| . \tag{1.2}$$

This integer N_ω is the index of f(z) at ω. If the indices N_ω are all bounded above for varying ω, then f is said to be of bounded index (b.i.) and the least such upper bound is the index of f. If $\sup_{\omega\in\mathbf{C}}\{N_\omega\} = \infty$, then f will be called of unbounded index. Now replacing ω by z in (1.2) we have

Definition 1.1. An entire function f is said to be of bounded index if there exists a nonnegative integer N such that

$$\max_{0\leqslant k\leqslant N}\left\{\frac{|f^{(k)}(z)|}{k!}\right\} \geqslant \frac{|f^{(n)}(z)|}{n!}, \qquad (f^{(0)} = f), \tag{1.3}$$

for all n and all z. The least such integer N is called the index of f.

Lepson first introduced the concept of functions of bounded index in connection with generalizations of hyperdirichlet series (see [43,44]). In his definition, Lepson uses strict inequality in (1.3). However, both definitions are equivalent (up to indices; see Theorem 4.1 and also Section 8). We follow Definition 1.1.

The functions sin z, cos z are of b.i. one, and a polynomial of degree n is of b.i., index not exceeding n. Functions with zeros of arbitrarily large multiplicity are of unbounded index but, as we will see below, there are functions of unbounded index and having simple zeros or no zeros.

2. Rate of Growth. An entire function of b.i. is of exponential type. We give two proofs of this fundamental result. First we introduce some notation. For any entire function f, let

$$M(r,f^{(j)}) = \max_{|z|=r} |f^{(j)}(z)|, \qquad T = \limsup_{r\to\infty} \frac{\log M(r,f)}{r}.$$

Invited survey article: Editors.

Further, let $\nu(r,f)$ denote the central index of f (see Section 8) and let

$$\gamma = \limsup_{r\to\infty} \frac{\nu(r,f)}{r} .$$

The function f is said to be of exponential type if $T < \infty$. Thus functions of mean type order one, of order less than one and all polynomials are included here in the class of functions of exponential type [1, p. 8].

Theorem 2.1. *If* f *is an entire function of bounded index, then*

$$\limsup_{r\to\infty} \frac{\log M(r,f)}{r} \leqslant N+1 , \tag{2.1}$$

where N *is the index of* f. *This result is sharp.*

We give two proofs of (2.1).

(a) (Hayman [31]). *Let* f *be an entire function of b.i.* N. *Then*

$$|f(z)| \leqslant \left\{ \max_{0\leqslant k\leqslant N} \left(\frac{|f^{(k)}(0)|}{(N+1)^k} \right) \right\} \exp\left((N+1)|z|\right) . \tag{2.2}$$

For the proof of (2.2) we require the following lemma which is also of independent interest. In this lemma the function f and the derivatives $f^{(k)}$ satisfy a "normalizing" condition different from (1.3).

Lemma 2.2. *Let* f *be entire and define*

$$f_n(z) = \max_{0\leqslant k\leqslant n-1} |f^{(k)}(z)| . \tag{2.3}$$

If

$$|f^{(n)}(z)| \leqslant f_n(z) \tag{2.4}$$

for all z, *then*

$$|f(z)| \leqslant f_n(0)\exp(|z|) . \tag{2.5}$$

Proof. Set, for a fixed real θ and $R > 0$,

$$g(t) = \max_{0\leqslant\nu\leqslant n-1} \{|f^{(\nu)}(te^{i\theta})|\}, \qquad 0 \leqslant t < R .$$

Clearly $g(t)$ is continuous and piecewise differentiable in $(0,R)$. In fact,

$$g'(t) \leqslant \max_{0\leqslant\nu\leqslant n-1} \{|f^{(\nu+1)}(te^{i\theta})|\} \leqslant \max\{g(t), |f^{(n)}(te^{i\theta})|\} .$$

The condition (2.4) now gives $g'(t) \leqslant g(t)$. Hence $e^{-t}g(t)$ is non-increasing in $[0,R)$. Setting $z = te^{i\theta}$ we get

$$f_n(z) = \max_{0\leqslant\nu\leqslant n-1} \{|f^{(\nu)}(z)|\} \leqslant g(0)e^{|z|} .$$

Since R is arbitrary and $|f(z)| \leqslant f_n(z)$, (2.5) follows.

Proof of (2.2). We have

$$\frac{|f^{(N+1)}(z)|}{(N+1)!} \leqslant \max_{0 \leqslant k \leqslant N} \left\{ \frac{|f^{(k)}(z)|}{k!} \right\} . \tag{2.6}$$

Set $F(z) = f(z/(N+1))$. Then from (2.6) we have

$$|F^{(N+1)}(z)| \leqslant \frac{(N+1)!}{(N+1)^{N+1-k}} \left\{ \max_{0 \leqslant k \leqslant N} \left(\frac{|F^{(k)}(z)|}{k!} \right) \right\}$$

$$\leqslant \max_{0 \leqslant k \leqslant N} \{|F^{(k)}(z)|\} .$$

Hence $F(z)$ satisfies (2.4) with $n = N+1$ and so we deduce from (2.5) that

$$|F(z)| \leqslant F_{N+1}(0)\exp(|z|)$$

$$= \left\{ \max_{0 \leqslant k \leqslant N} \left(\frac{|f^{(k)}(0)|}{(N+1)^k} \right) \right\} \exp(|z|) .$$

Since $F(z) = f(z/(N+1))$, (2.2) follows.

(b) (Shah [53]). *Let* f *be an entire function of b.i.* N. *Then*

$$\limsup_{r \to \infty} \frac{\nu(r,f)}{r} \leqslant N+1 . \tag{2.7}$$

For the proof we require three lemmas which we state without proofs. See [53] and the references given there.

Lemma 2.3. *Let* $F(r)$ *be positive and non-decreasing for* $r > r_0$ *and suppose that*

$$\limsup_{r \to \infty} \frac{F(r)}{r} = a , \qquad 0 < a \leqslant \infty ;$$

then corresponding to each pair of positive numbers b, c *satisfying the inequalities* $b < a$, $1 < c < a/b$, *there is a sequence* $R_1, R_2, \ldots$ *tending to* ∞ *such that* $R_1 > r_0$ *and*

$$F(r) > br , \qquad R_n \leqslant r \leqslant cR_n < R_{n+1} ; \qquad n = 1,2, \ldots .$$

Lemma 2.4. *For* $j = 1,2, \ldots$

$$\left(\frac{\nu(r,f)}{r}\right)^j M(r,f)(1-\epsilon_j(r)) \leqslant M(r,f^{(j)}) \leqslant \left(\frac{\nu(r,f)}{r}\right)^j M(r,f)(1+\epsilon_j(r)) , \tag{2.8}$$

where $0 \leqslant \epsilon_j(r) \to 0$, *for* $r \to \infty$ *outside a set of finite logarithmic measure.*

Lemma 2.5. *Let* f *be an entire function of positive order* ρ. *Then*

$$\rho \limsup_{r \to \infty} \frac{\log M(r,f)}{r^\rho} \leqslant \limsup_{r \to \infty} \frac{\nu(r,f)}{r^\rho} \leqslant e\rho \limsup_{r \to \infty} \frac{\log M(r,f)}{r^\rho} .$$

Proof of (2.7). We may assume $\gamma = \limsup \frac{\nu(r,f)}{r} > 0$, for otherwise (2.7) is obvious. Take $F(r) = \nu(r,f)$, $a = \gamma$, and b and c as in Lemma 2.3 and denote by E_n the interval $[R_n, cR_n]$. Write $G = \cup_{n=1}^{\infty} E_n$. The variation of $\log r$ in E_n is $\log c$ and so the total variation of $\log r$ over $\cup_1^p E_n$ tends to infinity with p. Since G is not of finite logarithmic measure there exists a sequence $\{r_n\}_1^\infty$ with $r_n \uparrow \infty$ and $r_n \in G$ such that (2.8) holds for $j = 1,2,\ldots,N+1$; that is,

$$\frac{M(r,f^{(j)})}{M(r,f)} \sim \left(\frac{\nu(r,f)}{r}\right)^j, \qquad r = r_n \to \infty. \tag{2.9}$$

By Lemma 2.3 we have $\nu(r_n,f) > br_n$ for $n = 1,2,\ldots$. Let $0 < b_1 < b$. Then by (2.9) we have for n sufficiently large

$$\frac{M(r_n,f^{(j+1)})}{M(r_n,f^{(j)})} > b_1 \qquad \text{for } j = 0,1,2,\ldots,N.$$

Suppose $b_1 \geqslant N+1$. Then

$$\frac{M(r_n,f^{(N+1)})}{(N+1)!} > \frac{M(r_n,f^{(N)})}{N!} > \cdots > M(r_n,f).$$

But since f is of index N we must have

$$\frac{M(r_n,f^{(N+1)})}{(N+1)!} \leqslant \max_{0\leqslant k\leqslant N} \left\{\frac{M(r_n,f^{(k)})}{k!}\right\},$$

and hence $b_1 < N+1$. Because $b_1 < b < a$ $(= \gamma)$ were chosen arbitrarily we must also have $\gamma \leqslant N+1$.

Now $T \leqslant \gamma$ and consequently each of these two inequalities (2.2) and (2.7) implies (2.1).

To show that (2.1) is sharp, we consider $f(z) = \exp((N+1)z)$ where N is a nonnegative integer. This function is of index N and type $T = N+1$.

3. Functions of unbounded index. Functions f of unbounded index must be transcendental functions, that is, functions f for which $\lim_{r\to\infty} \log M(r,f)/\log r = \infty$, for a polynomial is of b.i. Except for this restriction a function f of unbounded index can have arbitrary growth. In fact, we have

Theorem 3.1 (Pugh and Shah [50]). *Let* f *be any transcendental entire function. It is always possible to find an entire function* g *of unbounded index such that*

$$\log M(r,f) \sim \log M(r,g) \qquad (r \to \infty). \tag{3.1}$$

The proof depends on a theorem of Edrei and Fuchs. Choosing f, in Theorem 3.1, to be of b.i., we see that it is always possible to find functions of unbounded index with the same asymptotic behavior as that of a function f of bounded index. We can also construct functions of unbounded index and satisfying not an asymptotic equality such as (3.1), but an inequality for all r.

Theorem 3.2 (Shah [53]; Lepson [43]). *Given any function* $\varphi(r)$ *such that* $\varphi(r) \geqslant c > 0$ *for* $r \geqslant 0$ *and* $\log\varphi(r)/\log r \to \infty$ *as* $r \to \infty$, *there exists an entire function* f *of unbounded index such that for every* $r \geqslant 0$, $M(r,f) \leqslant \varphi(r)$.

If a function f has zeros of arbitrarily large multiplicity or if it is not of exponential type, then f is of unbounded index. But we also have functions such that they are of exponential type and have only simple zeros and are of unbounded index.

Theorem 3.3 (Shah [55]). *Let* $a \geq 0$, $a_1 \geq 1$ *and*

$$a_{k+1} \geq \max \{3(k+1)a_k, a_k^{(k+1)/k}\}, \qquad k = 1,2,\ldots, \tag{3.2}$$

and let

$$f(z) = e^{az} \prod_{m=1}^{\infty} \left(1 - \frac{z}{a_m}\right)^m . \tag{3.3}$$

Then $f(z)$ *is a transcendental entire function and the function*

$$F(z) = f(z) - c, \qquad \operatorname{Im} c \neq 0, \tag{3.4}$$

has all simple zeros, is of exponential type and is of unbounded index.

For the proof we require the following

Lemma 3.4. *Let*

$$P(z) = \prod_{n=1}^{\infty} \left(1 - \frac{z}{a_n}\right)^{k_n}$$

where $\{k_n\}_{n=1}^{\infty}$ *is any sequence of positive integers and* $\{a_n\}_{n=1}^{\infty}$ *is any strictly increasing sequence of positive numbers such that* $\Sigma_{n=1}^{\infty} k_n/a_n < \infty$. *Then* $P(z)$ *is a transcendental entire function and the zeros of* $e^{az}P(z) - c$ *are all simple. Here* a *and* c *are constants,* a *is real and* $\operatorname{Im} c \neq 0$.

Proof. By Laguerre's theorem [62; 52, p. 314] the zeros of $\frac{d}{dz}\{e^{az}P(z)\} = e^{az}(aP(z) + P'(z))$ are all real. Consider now

$$F(z) = e^{az}P(z) - c, \qquad \operatorname{Im} c \neq 0 .$$

Then $F(z)$ has no multiple zeros. For if ξ be a multiple zero, then we must have

$$F'(\xi) = e^{a\xi}(aP(\xi) + P'(\xi)) = 0$$

and thus ξ must be real. Clearly $e^{a\xi}P(\xi)$ is real and therefore $\operatorname{Im} F(\xi) = -\operatorname{Im} c \neq 0$ which contradicts that ξ is a zero of $F(z)$. Hence F has all simple zeros.

Proof of Theorem. It is easily seen that f is an entire function of order one and type a if $a \neq 0$, and of zero order if $a = 0$. By Lemma 3.4, the zeros of $f(z) - c$ are all simple. Further,

$$f(z) = \frac{f^{(n)}(a_n)}{n!}(z - a_n)^n + \frac{f^{(n+1)}(a_n)}{(n+1)!}(z - a_n)^{n+1} + \cdots .$$

Hence

$$\frac{f^{(n)}(a_n)}{n!} = \lim_{z \to a_n} \left\{ \frac{e^{az} \prod_{m=1}^{n-1} \left(1 - \frac{z}{a_m}\right)^m \left(1 - \frac{z}{a_n}\right)^n \prod_{m=n+1}^{\infty} \left(1 - \frac{z}{a_m}\right)^m}{(z - a_n)^n} \right\}$$

and thus,

$$\frac{|f^{(n)}(a_n)|}{n!} = e^{aa_n}\left|\prod_{m=1}^{n-1}\left(1-\frac{a_n}{a_m}\right)^m \frac{1}{a_n^n}\prod_{m=n+1}^{\infty}\left(1-\frac{a_n}{a_m}\right)^m\right| .$$

Using (3.1) we obtain

$$\left|\prod_{m=n+1}^{\infty}\left(1-\frac{a_n}{a_m}\right)^m\right| > \frac{1}{3} ,$$

$$\left|\prod_{m=1}^{n-1}\left(1-\frac{a_n}{a_m}\right)^m\right|\frac{1}{a_n^n} > e^{-1/3}(1+o(1))\exp\left\{\frac{n^2-9n+8}{2(n-3)}\log a_{n-3}\right\} .$$

Hence $|f^{(n)}(a_n)|/n! \to \infty$ as $n \to \infty$, whether $a > 0$ or $a = 0$. Choose n_0 such that

$$\frac{|f^{(n)}(a_n)|}{n!} > |c| \qquad \text{for all } n \geq n_0 .$$

Then for $n \geq n_0$ and $z = a_{n+1}$

$$\max\left\{|f(z)-c|, \frac{|f'(z)|}{1!}, \ldots, \frac{|f^{(n)}(z)|}{n!}\right\} = |c| < \frac{|f^{(n+1)}(z)|}{(n+1)!} .$$

Hence $F(z) = f(z) - c$ is of unbounded index, and the proof is complete.

Let $D(a,r) = \{z \mid |z-a| < r\}$ ($a \in \mathbf{C}$, $r > 0$). The above argument shows that for the function f given by (3.3), $|f(a_n + e^{i\theta})| > |c|$ ($n > n_0(c)$). We now use Rouche's theorem [52, p. 155] and get that $F(z) = f(z) - c$ has the same number of zeros in $D(a_n,1)$ as $f(z)$ has, that is n. This result should be contrasted with the following theorem on functions of b.i.

Theorem 3.5 (Hayman [31], Fricke [21]). *The number of zeros of a function* f *of b.i. in any disc* $D(a,1)$ *for varying* a *is bounded.*

Before we give some more theorems of this type we define the class of functions of "bounded value distribution," a concept due to Turan (see [29, p. 17]).

Consider functions f holomorphic in the disc $D(a,r)$. Let $n(a,r,\frac{1}{f-w})$ denote the number of zeros of $f-w$ ($w \in \mathbf{C}$, $a \in \mathbf{C}$) in $D(a,r)$, multiple zeros counted in accordance with their multiplicities. If $n(a,r,\frac{1}{f-w})$ is bounded for varying w, then we denote the least upper bound of $n(a,r,\frac{1}{f-w})$ by $p(a,r)$. This integer $p(a,r)$ is called the valency of f in $D(a,r)$. Writing $p = p(a,r)$ we say that f is p-valent in the disc $D(a,r)$. The valency of f in $D(a,r)$ is infinite if $n(a,r,\frac{1}{f-w})$, for varying w, is unbounded.

When f is entire, r can have any value in $(0,\infty)$. If $p(a,r)$ is bounded for varying a, we denote by $P(r)$ the least upper bound of the valencies $p(a,r)$ of f. We take $P(r)$ equal to ∞ if $p(a,r)$ is unbounded for varying a.

Definition 3.6. An entire function f is said to be of bounded value distribution (b.v.d.) if for every $R > 0$, there exists a fixed integer $P(R) > 0$ such that

(3.5) $$n(a,R,\frac{1}{f-w}) \leqslant P(R)$$

for all a and all w.

The function F defined by (3.4) is of unbounded value distribution. Note that F' is of unbounded index.

We now give some theorems of Hayman relating the functions of b.i., functions of b.v.d. and p-valent functions.

Let $A, A_1, \ldots$ denote positive constants.

Theorem 3.7 (Hayman [31]). *If* f *is p-valent in the disc* D(a,R) *for every* a *and a fixed* $R > 0$, *then* f' *has index at most* Ap max (1, 1/R).

Theorem 3.8 (Hayman [31]). *Suppose that* f(z) *is regular in* D(a,R) *and satisfies there*

$$(CR)^{p+1}\left|\frac{f^{(p+1)}(z)}{(p+1)!}\right| \leqslant \max_{1 \leqslant \nu \leqslant p}\left\{(CR)^{\nu}\frac{|f^{(\nu)}(z)|}{\nu!}\right\}$$

with $C \leqslant \frac{1}{2}$. *Then* f(z) *is p-valent in* $D(a, CR/\{12.2(p+1)^{\frac{1}{2}}\})$.

Theorem 3.9 (Hayman [31]). *An entire function* f *has b.v.d. if and only if* f' *has bounded index. More particularly, if* N *is the index of* f' *we have*

$$P\left\{\frac{1}{12.2}(N+2)^{-\frac{1}{2}}\right\} \leqslant N+1 \leqslant P(A_1^N) .$$

This theorem implies that if f has b.v.d. then

(3.6) $$\limsup_{r\to\infty} \frac{\log M(r,f)}{r} \leqslant N+1 ,$$

where N is the index of f'.

Theorem 3.10 (Hayman [31]). *Suppose that* f' *is a function of b.i.* N. *Then*

$$A_2(N+1) \leqslant P(1) \leqslant A_3(N+1) .$$

Furthermore, for $R \geqslant 1$, *we have*

$$P(R) < (N+1)e(R+2) .$$

Corollary 3.11. *If* $0 < r_1 < r_2 < \infty$ *then we have*

$$\frac{P(r_2)}{r_2} < A_4\frac{P(r_1)}{r_1} .$$

We note that if f' has b.i. then f has b.i., but there exist functions f having b.i. and such that f' has unbounded index. We return to this topic in Section 7.

4. Functions of bounded index. A characterization of the functions of b.i. can be stated as follows:

Theorem 4.1. *An entire function* f *is of b.i. if and only if there exists an integer* $N > 0$ *and a constant* $C > 0$ *such that*

$$|f^{(N+1)}(z)| \leq C \max_{0 \leq k \leq N} \{|f^{(k)}(z)|\} \tag{4.1}$$

for all z (cf. [31]).

If $C = 1$, we get (2.4) with n replaced by $N+1$. The proof of (4.1) is similar to that of (2.2). It is obvious that the integer N in (4.1) may not be the index of f. It will depend on C.

Theorem 4.2 (Fricke [21]). *An entire function* f *is of b.i. if and only if there exist a constant* $C > 0$ *and an integer* $N \geq 0$ *such that*

$$\sum_{j=0}^{N} \frac{|f^{(j)}(z)|}{j!} > C \sum_{j=N+1}^{\infty} \frac{|f^{(j)}(z)|}{j!} \tag{4.2}$$

for all z.

Functions f satisfying conditions similar to (4.2) and with $f^{(j)}(z)$ replaced by $M(r,f^{(j)})$ or by $I(r,j)$, where $I(r,\ell) = \{\int_0^{2\pi} |f^{(\ell)}(re^{i\theta})|^p \, d\theta\}^{1/p}$ $(p \geq 1)$ have been studied by Gross [29] and others (see [19,21, 61,64]). These functions are all of exponential type (see Theorems 1, 2, 3 of [61], Theorem 2 of [19] and Theorems 9 and 12 of [21]), but as we mentioned earlier there are functions of exponential type and of unbounded index. Consider, for instance,

$$f(z) = \prod_{n=1}^{\infty} \left(1 + \frac{z}{na^{n+1}}\right)^n, \qquad a \geq 2. \tag{4.3}$$

Then f is of order zero, satisfies a condition of the type (4.2) with $f^{(j)}(z)$ replaced by $M(r,f^{(j)})$ and is not of b.i., but it is of M-bounded index defined below.

Definition 4.3. An entire function f is said to be of M-bounded index if there exists an integer N such that

$$\max_{0 \leq j \leq N} \left\{\frac{M(r,f^{(j)})}{j!}\right\} \geq \frac{M(r,f^{(n)})}{n!} \tag{4.4}$$

for all $r \geq 0$ and all n. The least such integer N is called the M-index of f.

It is easy to see that an entire function of index N is also of M-index (we write it as N(M)) not exceeding N. For $f(z) = ze^z$, we have $N = 2$, $N(M) = 1$.

Theorem 4.4 (Fricke, Shah and Sisarcick [19]). *An entire function* f *is of exponential type if and only if there exists an integer* $N \geq 0$ *such that* f *is of M-bounded index not exceeding* N. *If* f *is of exponential type* T, *then* $T \leq N(M) + 1$. *This upper bound on* T *is sharp.*

The class of functions of b.i. are characterized by locally slow growth. Furthermore, this class is closed under translation and multiplication.

Theorem 4.5 (Fricke [21]). *An entire function* f *is of b.i. if and only if for each* $r > 0$, *there exists* $M > 0$ *such that for all* $z \in C$

$$\max_{|\xi-z|=2r} \{|f(\xi)|\} \leq M \max_{|\xi-z|=r} \{|f(\xi)|\} .$$

This theorem enables one to prove that if f is of b.i. then f is of exponential type.

Theorem 4.6 (Fricke [16]). *If* $f(z)$ *is of b.i. then* $g(z) = f(az+b)$ *is of b.i. for any* $a,b \in C$.

Theorem 4.7 (Fricke [16,18]). (i) *If* f *and* g *are entire functions of b.i. then* $h(z) = f(z)g(z)$ *is also of bounded index.*

(ii) *Let* f, g *be two entire functions and let* $h(z) = f(z)g(z)$. *If* h *and* f *are of b.i. then* g *must also be of bounded index.*

5. Functions of strongly bounded index.

In this section we define functions of strongly bounded index and consider the sum of two functions of b.i.

Let

$$\Omega(z) = \Omega_s(z,f) = \max_{0 \leq j \leq s} \left\{ \frac{|f^{(j)}(z)|}{j!} \right\} . \tag{5.1}$$

Definition 5.1. An entire function $f(z)$ is of strongly bounded index (we shall also say of the class SB) if there exist a number χ, $0 < \chi < 1$ and an integer $s \geq 0$ such that

$$\frac{|f^{(n)}(z)|}{n!} \leq \chi \Omega_s(z,f) \tag{5.2}$$

for all $n \geq s+1$ and all z.

For instance, $f(z) = e^z \in SB$. Here $\chi = ½$, $s = 1$.

We note that, by Theorem 4.6, the concepts of b.i. and strongly b.i. are identical (up to indices). This concept is useful when one is interested in knowing whether a function has b.i. and not so much the index (see also Theorem 4.1). We have

Theorem 5.2 (Shah and Shah [60]). *Let* $f(z) \in SB$. *Then*

(i) $f(z)$ *is of b.i.*;

(ii) *if* $P(z)$ *is a polynomial then* $f(z)P(z) \in SB$;

(iii) $\{f(z)/P(z)\} \in SB$ *provided* $\{f(z)/P(z)\}$ *is entire*;

(iv) *if* $a \in C$ *and* $0 < \chi < e^{-2|a|}$,

where χ *is the constant in* (5.2), *then* $e^{az}f(z) \in SB$.

The next two theorems show that the class of functions of b.i. is not closed under addition.

Theorem 5.3 (Pugh [49]). *Let* $g(z) \in SB$. *Let* $f(z)$ *be any entire function such that*

$$f(0) \neq 0 , \tag{5.3}$$

(5.4) $$M(2|z|,f) \leqslant \Omega_s(z,g) \qquad \text{for all } z\,.$$

Then if the constant $c \neq 0$ *is chosen small enough, the function*

(5.5) $$h(z) = g(z) + cf(z)$$

is of index not greater than s.

Proof. By Cauchy inequality

(5.6) $$\frac{|f^{(n)}(z)|}{n!}\,\Gamma^n \leqslant M(2\Gamma,f)\,, \qquad |z| = \Gamma$$

for all z and all integers $n \geqslant 0$. From (5.6) we deduce

(5.7) $$\frac{|f^{(n)}(z)|}{n!} \leqslant BM(2\Gamma,f)\,, \qquad |z| = \Gamma\,,$$

where $B = M(2,f)/|f(0)|$, for all z and all integers $n \geqslant 0$. Write $\Omega_s(z,g) = \Omega(z)$. Differentiating h defined by (5.5), we have

(5.8) $$\frac{|h^{(n)}(z)|}{n!} \leqslant \chi\Omega(z) + |c|BM(2\Gamma,f) \leqslant (\chi + |c|B)\Omega(z)$$

for $n \geqslant s+1$ and all z. If $n \leqslant s$ we have

$$\frac{|h^{(n)}(z)|}{n!} \geqslant \frac{|g^{(n)}(z)|}{n!} - |c|\,\frac{|f^{(n)}(z)|}{n!} \geqslant \frac{|g^{(n)}(z)|}{n!} - |c|BM(2\Gamma,f)\,,$$

$0 \leqslant n \leqslant s$. Hence

(5.9) $$\max_{0\leqslant n\leqslant s}\left\{\frac{|h^{(n)}(z)|}{n!}\right\} \geqslant \Omega(z)(1-|c|B)\,.$$

Now choose $c \neq 0$ such that

$$|c|B < 1\,, \qquad \frac{\chi + |c|B}{1-|c|B} < 1\,.$$

Then (5.8) and (5.9) yield

$$\frac{|h^{(n)}(z)|}{n!} \leqslant \max_{0\leqslant k\leqslant x}\left\{\frac{|h^{(k)}(z)|}{k!}\right\}$$

and so $h(z)$ is of b.i. not exceeding s.

This theorem enables one to construct functions of b.i. such that their sum is of unbounded index.

Theorem 5.4 (Pugh [49]). *The sum of two functions of b.i. need not be of bounded index.*

Proof. Let $g(z) = \cos z + \cosh z$. Then $g^{(4)}(z) = g(z)$. Hence if we set

$$\max_{0\leqslant j\leqslant 3}\left\{\frac{|g^{(j)}(z)|}{j!}\right\} = \Omega(z)\,,$$

we obtain

$$\frac{|g^{(n)}(z)|}{n!} \leqslant \frac{\Omega(z)}{4}\,, \qquad (n \geqslant 4)\,.$$

Now it can be shown that, for all z,

$$\Omega(z) \geqslant A \exp(|z|/2), \qquad (0 < A = \text{const.}).$$

Let

(5.10) $$\varphi(z) = \prod_{j=1}^{\infty} \left(1 + \frac{z}{e^j}\right)^j .$$

Then $\varphi(z)$ is of unbounded index and of zero order. Hence

$$M(2\Gamma,\varphi)/\Omega(z) \leqslant L, \qquad |z| = \Gamma$$

where L is a finite bound. Then the function

$$F(z) = cf(z) = \frac{c}{L}\varphi(z)$$

is of unbounded index and $h(z) - g(z) = cf(z) = F(z)$ where $h(z)$ and $g(z)$ are functions of bounded index.

We shall see later (Theorem 7.2) that there exists a function f of unbounded index such that $F(z) = f(z) - c$, where c is any non-zero complex number, is of bounded index.

In Theorem 2.1 we saw that a function of b.i. is of exponential type but the converse is not true. See, for instance, the function in (5.10). But we can prove the following result with the help of Theorem 5.3.

Theorem 5.5 (Shah [63])[1]. *An entire function of exponential type can be expressed as the difference of two entire functions of bounded index.*

Proof. Let $W(z)$ be the given function and write

$$\limsup_{r\to\infty} \frac{\log M(r,W)}{r} = T < \infty .$$

Assume first $W(0) \neq 0$. Given $\epsilon > 0$, we can find r_0 such that

$$M(2r,W) < \exp\{2r(T+\epsilon)\}, \qquad r \geqslant r_0 .$$

Let $\delta = 4(T+\epsilon)+1$, $K = \max_{|z|=2r_0} |W(z)|$. We now define

(5.11) $$g(z) = K(\cos \delta z + \cosh \delta z).$$

Then $g(z)$ satisfies the differential equation

(5.12) $$g^{(4)}(z) = \delta^4 g(z);$$

and we have

$$\frac{g^{(n)}(z)}{n!} = \frac{\delta^{n-k}k!}{n!}\left\{\frac{g^{(k)}(z)}{k!}\right\}$$

---

[1] I am grateful to Dr. S. N. Shah for a preprint of his paper. Walter Pugh (now deceased) and S. N. Shah were students of Professor Albert Edrei.

where $k \equiv n \pmod 4$, $0 \leqslant k \leqslant 3$. Let N be the smallest integer $\geqslant 3$ such that

$$\frac{6\delta^N}{N!} \leqslant \frac{1}{2}$$

and let

$$\Omega(z) = \max_{0 \leqslant j \leqslant N} \{|g^{(j)}(z)|/j!\} . \tag{5.13}$$

Then for all $n \geqslant N+1$ and all z

$$\frac{|g^{(n)}(z)|}{n!} = \frac{\delta^{n-k}k!}{n!} \left\{\frac{|g^{(k)}(z)|}{k!}\right\} \leqslant \tfrac{1}{2}\Omega(z) .$$

This means that $g(z) \in SB$ and therefore it is of b.i. by Theorem 5.2. Further, if g and Ω are defined by (5.11) and (5.13), then it can be shown that for all z

$$\Omega(z) \geqslant B \exp(\delta|z|/2) \tag{5.14}$$

where $B > 0$ is a suitably chosen constant. Hence for $|z| = r$,

$$\Omega(z) \geqslant B \exp\{\delta|z|/2\} > B \exp\{2(T+\epsilon)r\} > BM(2r,W) \qquad (r \geqslant r_0) .$$

For $r < r_0$

$$\Omega(z) \geqslant B\frac{M(2r,W)}{M(2r_0,W)} ,$$

and therefore setting $L = \{1 + M(2r_0,W)\}/B$ we have for all $r \geqslant 0$

$$M(2r,W) \leqslant L\Omega(z) , \qquad |z| = r .$$

Hence the function $f(z) = W(z)/L$ satisfies conditions (5.3) and (5.4) and therefore if $c \neq 0$ is small enough

$$p(z) = g(z) + cf(z) = g(z) + \frac{c}{L} W(z) \tag{5.15}$$

is of b.i. Hence

$$W(z) = \frac{L}{c} p(z) - \frac{L}{c} g(z)$$

is the required representation if $W(0) \neq 0$. We now consider the case $W(0) = 0$. Let $W_1(z) = W(z) + 1$. Then by the above case

$$W_1(z) = \frac{L}{c} p(z) - \frac{L}{c} g(z)$$

where g defined by (5.11) is of b.i. and $p(z)$ is of b.i. by Theorem 5.3. Write

$$G(z) = g(z) + \frac{c}{L} .$$

Then $G^{(5)}(z) = \delta^4 G'(z)$ and hence ([53] ; see Theorem 9.1 below) $G(z)$ is a function of b.i. Consequently,

$$W(z) = \frac{L}{c} p(z) - \frac{L}{c} G(z)$$

gives, in the case $W(0) = 0$, the representation as the difference of functions of bounded index.

6. Spacing of zeros. We saw, in Theorem 3.5, that the number of zeros of a function of b.i. in any disc of radius one has an upper bound. Theorems 3.9 and 3.10 imply that f' is of b.i. if and only if there exists an integer $p > 0$ such that f is at most p-valent in any disc of radius one (see also [17]). By placing some conditions on the spacing of the zeros of a function of exponential type we obtain functions of bounded index.

Theorem 6.1 (Pugh and Shah [50]). *Let* f *be an entire function defined by*

$$f(z) = \prod_{n=1}^{\infty} \left(1 - \frac{z}{a_n}\right), \tag{6.1}$$

and suppose that

$$|a_1| \geqslant a = 5, \qquad |a_{n+1}| \geqslant a^n |a_n|, \qquad (n = 1,2,\ldots). \tag{6.2}$$

Then for all z

$$|f^{(n)}(z)| \leqslant \max\{|f(z)|, |f'(z)|\}, \qquad (n = 2,3,\ldots), \tag{6.3}$$

and

$$f(z), f'(z), f''(z), \ldots \text{ are each of b.i. one}. \tag{6.4}$$

If a function is defined, as in (6.1) and (6.2), then we will say that the function has widely spaced zeros. To prove the theorem we require four lemmas.

Lemma 6.2. *Let* f *be defined by* (6.1) *and* (6.2). *Let* $\{b_j\}_{j=1}^{\infty}$ $(|b_j| \leqslant |b_{j+1}|)$ *be the zeros of* f'. *Then*

$$\frac{|a_{n+1}|}{b} < |b_n| \leqslant |a_{n+1}|, \qquad (n \geqslant 2,\ b = 1.6), \tag{6.5}$$

and

$$\left(1 + \frac{2R + d}{a}\right)|a_1| < |b_1| \leqslant |a_2|, \qquad (R = 2.4,\ d = 10^{-3},\ |a_1| \geqslant a = 5). \tag{6.6}$$

To prove this lemma we use Rouche's theorem and the theorem of Gauss-Lucas.

Lemma 6.3. *If* f *has widely spaced zeros, all the derivatives* $f', f'', \ldots$ *have the same property.*

Proof. It is sufficient to prove that if f has widely spaced zeros, the zeros of f' are also widely spaced. Write $1 + (2R + d)/a = 1.9602 = c$. Then by (6.6)

$$9.801 \leqslant c|a_1| < |b_1|. \tag{6.7}$$

By (6.5) and (6.6)

$$|b_n| \leqslant |a_{n+1}|, \qquad \frac{1}{b}|a_{n+2}| < |b_{n+1}|, \qquad (n \geqslant 1).$$

Hence

(6.8) $$\left|\frac{b_{n+1}}{b_n}\right| > \frac{|a_{n+2}|}{b|a_{n+1}|} \geqslant \frac{a^{n+1}}{b} > a^n , \qquad (n \geqslant 1) .$$

The relations (6.7) and (6.8) show that the b's are widely spaced.

We now state two lemmas without proof (see [50]). Lemma 6.5 implies that

$$\inf_{\substack{1 \leqslant j < \infty \\ 1 \leqslant k < \infty}} |a_j - b_k| > 2R + d .$$

Denote the zeros of $f^{(k)}$, in order of ascending moduli, by $\{a_j^{(k)}\}_{j=1}^{\infty}$. Then $a_n^{(0)} = a_n$. Let

$$D_k(\rho) = \bigcup_{j=1}^{\infty} \{z\colon |z - a_j^{(k)}| \leqslant \rho\} , \qquad (\rho > 0, k = 0,1,2, \ldots) .$$

Lemma 6.4. *If* f *has widely spaced zeros and if* $z \notin D_0(R)$, *then*

$$\left|\frac{f'(z)}{f(z)}\right| < \frac{\sqrt{2}}{2} , \qquad \left|\frac{f''(z)}{f(z)}\right| < 1 .$$

Lemma 6.5. *If* $z \in D_0(2R + d)$ *then* $f'(z) \neq 0$.

Proof of Theorem. Because all the derivatives of f have widely spaced zeros, Lemmas 6.2 to 6.5 apply to all of the functions $f^{(k)}$ $(k = 0,1, \ldots)$. In particular, Lemma 6.5 shows that the sets $D_{n-2}(R)$ and $D_{n-1}(R)$ are disjoint for $n \geqslant 2$. Hence at least one of the two inequalities

$$\left|\frac{f^{(n)}(z)}{f^{(n-2)}(z)}\right| < 1 , \qquad \left|\frac{f^{(n)}(z)}{f^{(n-1)}(z)}\right| < 1 , \qquad (n \geqslant 2)$$

must hold. Thus for all z

(6.9) $$|f^{(n)}(z)| < \max\{|f^{(n-1)}(z)|, |f^{(n-2)}(z)|\} \qquad (n = 2,3, \ldots)$$

and (6.3) follows by an induction over n. To prove the remaining part, we use (6.9).

Corollary 6.6. *There exist functions of b.i. and of arbitrarily slow growth.*

In [34] Mrs. King has proved (6.3) when $a = 4$ in (6.2).

If the zeros a_n are positive then we can relax the condition (6.2).

Theorem 6.7 (Shah and Shah [60]). *Let* $\{a_n\}_{n=1}^{\infty}$ *be a positive strictly increasing sequence such that*

(6.10) $$a_{n+1} - a_n \geqslant d_n , \qquad (n \geqslant 1) ,$$

where $\{d_n\}_{n=1}^{\infty}$ *is positive non-decreasing and* $\Sigma_1^{\infty} 1/(nd_n) < \infty$. *Then*

(6.11) $$f(z) = \prod_{n=1}^{\infty} \left(1 - \frac{z}{a_n}\right) \in SB .$$

To prove this theorem we choose m suitably large and show that

$$F(z) = \prod_{j=m}^{\infty} \left(1 - \frac{z}{a_j}\right) \in SB .$$

We now use Theorem 5.2 (ii).

Corollary 6.8. *Let* f *be defined by* (6.11) *and suppose that*

$$\frac{a_{n+1}}{a_n} \geqslant c > 1 , \qquad a_1 > 0 .$$

Then $f \in SB$. *Furthermore, if* $F(z) = \exp(\alpha z + \beta) f(z)$, $(\alpha, \beta \in C)$, *then each* $F^{(k)}(z)$ $(k = 0,1,2, \ldots)$ *is of b.i.* (see [18]).

Corollary 6.9. *The Lindelof functions*

$$f(z) = \prod_{n=2}^{\infty} \left(1 - \frac{z}{a_n}\right)$$

where $a_n = \{n(\log n)^{\alpha}\}^{1/\lambda}$, $0 < \lambda \leqslant 1$ *and* $\alpha > 1$ *if* $\lambda = 1$, α *an arbitrary real number if* $0 < \lambda < 1$, *belong to the class* SB.

Corollary 6.10. *Let* $\{a_n\}_{n=1}^{\infty}$ *satisfy the conditions of Theorem 6.7, and let* $\{a_{n_j}\}_{j=1}^{\infty}$ *be any one of its infinite subsequences. Then*

$$g(z) = \prod_{j=1}^{\infty} \left(1 - \frac{z}{a_{n_j}}\right) \in SB .$$

If we choose in Corollary 6.10, a subsequence $\{a_{n_j}\}$ by omitting from the given sequence $\{a_n\}$ long sections of consecutive terms we get a function h (say) belonging to the class SB. Furthermore, by suitable choices of the gaps we can obtain irregularities in the growth of the function h.

Corollary 6.11. *Let* $0 \leqslant \lambda \leqslant \rho \leqslant 1$ *be given. Then there exists a function* h *of b.i. and of order* ρ *with lower order* λ.

A theorem analogous to Theorem 3.1 but for functions of exponential type is as follows. Let $n(r, 1/g)$ denote the number of zeros of the function g in $|z| \leqslant r$.

Theorem 6.12 (Fricke [25]). *Let* f *be an entire function of exponential type with order* ρ *and lower order* λ. *If* $\rho - \lambda < 1$ *then there exists an entire function* g *of b.i. such that*

$$\int_1^r \frac{n(t, 1/g)}{t}\, dt \sim \log M(r,g) \sim \log M(r,f) , \qquad (r \to \infty) .$$

The index of the Lindelof function f, when $\alpha = 0$ and $\lambda < 1$, has be considered in

Theorem 6.13 (King and Shah [35]). *The Lindelof function*

$$f(z) = f(z,\beta) = \prod_{n=1}^{\infty} \left(1 - \frac{z}{n^{\beta}}\right) , \qquad \beta > 1$$

is of bounded index. The index $N = 1$ *if* $\beta \geqslant 3$.

The function $f(z,2) = (\sin \pi\sqrt{z})/\pi\sqrt{z} = \prod_{n=1}^{\infty}\left(1 - \frac{z}{n^2}\right)$ is of b.i. $N > 1$. A similar theorem holds for Mittag-Leffler function of order $1/k$ when k is an integer [3, p. 50; 52, p. 345].

Theorem 6.14 (Shah [54]). *Let*

$$F(z) = F(z,k) = \sum_{n=0}^{\infty} \frac{z^n}{\Gamma(1+nk)} \qquad k = 1,2,\ldots.$$

Then $F(z)$ *is an entire function of bounded index.*

The condition of monotonicity on $\{d_n\}$ in Theorem 6.7 cannot be dropped for we have

Theorem 6.15 (Fricke [20]). *There exists an entire function*

$$f(z) = \prod_{n=1}^{\infty} \left(1 - \frac{z}{a_n}\right)$$

with $a_{n+1} > a_n > 0$ $(n \geq 1)$, $\Sigma \frac{1}{a_n} < \infty$, $\lim_{n\to\infty} d_n = \infty$, $\Sigma \frac{1}{nd_n} < \infty$ *where* $d_1 = a_1$, $d_n = a_n - a_{n-1}$ $(n > 1)$, *such that* f *is not of bounded index.*

A theorem giving a condition, when the zeros a_n are not necessarily real, is as follows.

Theorem 6.16 (Fricke [18]). *Let* f *be an entire function of exponential type. Then* f *is of b.i. if and only if for each* $d > 0$, *there exists* $M = M(d) > 0$ *such that* $|f'(z)| \leq M|f(z)|$ *for all* z *with* $|z - a_n| \geq d$ *for all* n, *where the* a_n*'s are the zeros of* f.

7. Derivatives of functions of b.i. We now show that the class of functions of b.i. is not closed under differentiation. In Theorem 7.1 we construct a function f of unbounded index and use this function in the next two theorems. Another example of a function of b.i. whose derivative is not of b.i. has been given by Hayman [31].

Theorem 7.1 (Shah [56]). *There exist strictly increasing sequences of positive numbers* $\{a_j\}_{j=1}^{\infty}$ *and positive integers* $\{k_j\}_{j=1}^{\infty}$ *such that the following four statements hold:*

(i) *The function* $f(z) = \Pi_{j=1}^{\infty} (1 - z/a_j)^{k_j}$ *is entire.*

(ii) *The discs*

$$\Delta_j = \left\{z\colon |z - a_j| \leq a_j\left(1 - \frac{1}{k_j}\right)\right\}$$

are disjoint. Let $\Delta = \cup_{j=1}^{\infty} \Delta_j$. *If* $\xi \in \Delta^c$ *(the complement of* Δ*) then*

$$|f(\xi)| \to \infty \qquad \text{and} \qquad \left|\frac{f^{(k)}(\xi)}{k!f(\xi)}\right| \to 0$$

uniformly in $k \geq 1$, *as* $\xi \to \infty$ *in* Δ^c.

(iii) *Let* $D_j = \{z\colon |z - a_j| \leq a_j/2\}$ *and* $D = \cup_{j=1}^{\infty} D_j$. *If* $\xi \in D$ *then*

$$\frac{|f^{(k)}(\xi)|}{k!} \to 0$$

uniformly in $k \geq 0$, *as* $\xi \to \infty$ *in* D.

(iv) *If* $\xi \in \Delta \backslash D$ *then we have either*

(a) $$|f(\xi)| \to \infty \qquad \text{and} \qquad \left|\frac{f^{(k)}(\xi)}{k!f(\xi)}\right| \to 0$$

uniformly in $k \geq 1$, *as* $\xi \to \infty$ *in* $\Delta \backslash D$; *or*

(b)
$$\left| \frac{f^{(k)}(\xi)}{k!\,f'(\xi)} \right| \to 0$$

uniformly in $k \geq 2$, *as* $\xi \to \infty$ *in* $\Delta \backslash D$.

We may take, for instance, $a_1 = k_1 = 10$, $k_{j+1} = k_j^2$ $(j \geq 1)$, $a_{j+1} = k_{j+1} \exp\{(\log \tfrac{3}{2})k_j\}$ $(j \geq 1)$. Then the discs Δ_j $(j = 1,2,\ldots)$ are all disjoint. For the proof see [56]. Theorems 7.2, 7.4 and Corollary 7.5 utilize these properties of this function f and the derivatives $f^{(k)}$.

Theorem 7.2 (Shah [56]). *There is an entire function* f *of unbounded index such that* $F(z) = f(z) - C$, *where* C *is any nonzero compelx number, is of bounded index.*

To prove this theorem, we use the following

Lemma 7.3. *Let* $F \not\equiv 0$ *be an entire function and* T *a given positive number. Then there exists a positive integer* P *such that for* $|z| \leq T$

$$\max_{0 \leq k \leq P} \left\{ \frac{|F^{(k)}(z)|}{k!} \right\} \geq \frac{|F^{(j)}(z)|}{j!}, \qquad j = 1,2,\ldots .$$

The proof is straightforward. See ([54, p. 133; 43, p. 305]).

Now let f be the function defined in Theorem 7.1 and $F(z) = f(z) - C$, where $C \neq 0$. Then f is of unbounded index. Further, it follows from Theorem 7.1 that there exists $T = T(C) > 0$ such that

$$\max \left\{ |F(z)|, \frac{|F'(z)|}{1!} \right\} \geq \frac{|F^{(j)}(z)|}{j!}, \qquad j = 2,3,\ldots$$

provided $|z| \geq T$. Hence, by the lemma, $F(z)$ is of bounded index.

Theorem 7.4 (Shah [56]). (i) *If* f *is any entire function such that* f' *is of b.i.* $N_{f'}$, *then* f *is also of b.i.* N_f *and* $N_f \leq N_{f'} + 1$.

(ii) *There exists an entire function* F *of b.i. such that* F' *is of unbounded index.*

Proof. (i) Write $N_{f'} = N$, $K(f') = \max\limits_{0 \leq p \leq N} \{|f^{(p+1)}(z)|/p!\}$. Then for $j \geq 2$,

$$\frac{|f^{(N+1)}(z)|}{(N+j)!} \leq \frac{|f^{(N+j)}(z)|}{(N+j-1)!}\,\frac{1}{N+2} \leq \frac{K(f')}{N+2} \leq \max_{0 \leq p \leq N+1} \left\{ \frac{|f^{(p)}(z)|}{p!} \right\}$$

and so $N_f \leq N + 1$. The inequality is sharp. Consider the function

$$f(z) = e^z - \left(1 + z + \cdots + \frac{z^N}{N!}\right).$$

Then $N_{f'} = N$, $N_f = N + 1$.

(ii) Let $F(z) = f(z) + 1$. Then F is of b.i. and $F' = f'$ is of unbounded index, since f' has zeros of arbitrarily large multiplicity. By (i) each $F^{(k)}$ $(k \geq 2)$ is of unbounded index.

Corollary 7.5. *There exist two entire functions* g *and* G *of b.v.d. such that* $h(z) = g(z) + G(z)$ *is not of b.v.d.*

Proof. Let $f(z)$ be the function defined in Theorem 7.1 and $\theta(z)$ be an entire function such that $\theta'(z) = f(z)$. Let $g(z) = \theta(z) + z$. Then $g'(z) = \theta'(z) + 1 = f(z) + 1$ and so by Theorems 7.2 and 3.9, g has b.v.d. Let $G(z) = -z$, $h(z) = g(z) + G(z) = g(z) - z$. Then G has b.v.d. and $h'(z) = g'(z) - 1 = f(z)$ is of unbounded index and so h is not of b.v.d.

This result should be compared with a theorem of Goodman [27] that there exist two functions g and G analytic and univalent in the disc $D(0,1)$ and both normalized $(g(0) = G(0), g'(0) = G'(0) = 1)$ such that the function $(g + G)/2$ has valence ∞ $D(0,1)$.

Note that the argument in Corollary 7.5 will not work with $g(z) = \theta(z) + c$. Unlike Theorem 7.2, if a function f has b.v.d. then so has $f+c$ for any complex number c. But we have the following results for functions of b.v.d. similar to those for functions of b.i.

Proposition 7.6. (i) *An entire function of b.v.d. is of exponential type but the converse is not true.*
(ii) *Every entire function of exponential type can be expressed as the difference of two entire functions of b.v.d.*

Proof. (i) For the first part see the remark following Theorem 3.9. For the second part, note that $\theta(z)$ of Corollary 7.5 is of exponential type but it is not of b.v.d.
(ii) Let $H(z)$ be a function of exponential type. Then, by Theorem 5.5, $H'(z) = h(z)$ can be expressed as the difference $g(z) - q(z)$ where g and q are entire functions of b.i. Hence

$$H(z) = \int_c^z g(w)\,dw - \int_c^z q(w)\,dw \qquad c \in \mathbf{C}$$

$$= G(z) - Q(z)$$

where G and Q are entire functions of b.v.d.

8. Growth problems and Index. Let f be an entire function of order ρ and lower order λ. Consider the power series expansion:

$$f(z) = \sum_{n=0}^{\infty} a_n z^n .$$

For $0 \leqslant r < \infty$, the sequence $\{|a_n| r^n\}_{n=0}^{\infty}$ converges to zero and hence there exists a largest nonnegative integer $\nu(r,f) = \nu(r)$ such that

$$\mu(r,f) = \max_{n \geqslant 0} |a_n| r^n = |a_{\nu(r)}| r^{\nu(r)} \geqslant |a_j| r^j \qquad \text{for } j = 0,1,2,\ldots .$$

$\mu(r,f)$ is called the maximum term and $\nu(r)$ the central index of f. We have ([48, pp. 3–13; 69])

$$\lim_{r\to\infty} \begin{matrix}\sup\\ \inf\end{matrix} \frac{\log \nu(r,f)}{\log r} = \lim_{r\to\infty} \begin{matrix}\sup\\ \inf\end{matrix} \frac{\log\log M(r)}{\log r} = \begin{cases}\rho\\ \lambda\end{cases} .$$

In a similar way, Frank [9] and Frank and Mues [11,12] introduce a function $I(r,f)$ to define a function of b.i. Consider the expansion (1.1) and let k_ω be the largest nonnegative integer such that

$$\frac{|f^{(k_\omega)}(\omega)|}{(k_\omega)!} \geqslant \frac{|f^{(n)}(\omega)|}{n!} \qquad \text{for } n = 0,1,2,\ldots$$

and let

$$I(r) = I(r,f) = \sup_{|\omega| \leqslant r} k_\omega .$$

Then $I(r,f)$ is a monotone increasing function of r.

Definition 8.1. If ([9])

$$\limsup_{r \to \infty} I(r) < \infty ,$$

then f is said to be of bounded index.

This definition is equivalent to Definition 1.1: if f is of b.i. according to Definition 1.1, then it is of b.i. according to Definition 8.1 and vice-versa.

Example 8.2. Let $f(z) = \exp((N+1)z)$, $N \geqslant 0$, integer. Here $\limsup_{r\to\infty} I(r) = N+1$. According to Frank and also Lepson, the index $N_f = N+1$ whereas according to Definition 1.1 (see also [27, 31, 16]) f has index N.

Example 8.3. Let $f(z) = e^{Q(z)}$ where $Q(z)$ is a polynomial of exact degree ρ, $\rho \geqslant 1$ and integer. We have

$$\lim_{r\to\infty} \frac{\log^+ I(r,f)}{\log r} = \rho - 1 . \tag{8.1}$$

In general we have

Theorem 8.4 (Frank and Mues [12]). *If f is an entire function of finite order ρ, and lower order λ, then*

$$\max(0,\rho-1) \leqslant \limsup_{r\to\infty} \frac{\log^+ I(r)}{\log r} \leqslant \rho , \tag{8.2}$$

$$\max(0,\lambda-1) \leqslant \liminf \frac{\log^+ I(r)}{\log r} . \tag{8.3}$$

The right hand inequality in (8.2) is sharp [9]. The left hand inequality in (8.2) is best possible when $\rho \geqslant 1$ is a rational number. In fact, we have

Theorem 8.5 (Frank and Mues [11]). *To every rational number $\rho \geqslant 1$, there corresponds an entire function f, of order ρ, and such that*

$$\lim_{r\to\infty} \frac{\log I(r,f)}{\log r} = \rho - 1 .$$

As in Definition (4.3) we consider replacing the derivatives $f^{(k)}(\omega)$ by the corresponding maximum moduli [12]. Let $J(r)$ be the largest integer such that

$$\frac{M(r,f^{J(r)})}{(J(r))!} \geqslant \frac{M(r,f^{(j)})}{j!} \qquad \text{for } j = 0,1,2,\ldots .$$

Then $J(r) \leqslant I(r)$ for all r and we have

Theorem 8.6 (Frank and Mues [12]). *If* f *is an entire function of finite order and lower order* λ, *then*

(8.4) $$\max(0,\rho-1) = \limsup_{r\to\infty} \frac{\overset{+}{\log} J(r)}{\log r}$$

(8.5) $$\max(0,\lambda-1) \leqslant \liminf_{r\to\infty} \frac{\overset{+}{\log} J(r)}{\log r} .$$

We saw earlier (Corollary 6.11; Theorem 6.12) that a function of b.i. can have any order ρ provided $0 \leqslant \rho \leqslant 1$. We now derive a condition so that a function of b.i. may have order one. Consider all indices N_ω defined by (1.2) and suppose that f is of b.i. N so that $N = \sup_\omega \{N_\omega\}$. The index set S_n of order n, $0 \leqslant n \leqslant N$, is by definition the set of all points ω such that $N_\omega = n$. Let $\sigma_n = \{r \mid |\omega| = r, \omega \in S_n\}$, $0 \leqslant n \leqslant N$ and let $m_\ell(\sigma_n)$ denote the logarithmic measure of the set $\sigma_n \cap [1,\infty)$.

Theorem 8.7 (Lee and Shah [39]). *Let* f(z) *be a transcendental entire function of b.i. If* $m_\ell(\sigma_n) = \infty$ *for some* $n > 0$, *then*

$$\limsup_{r\to\infty} \frac{\nu(r)}{r} \geqslant n \qquad \text{and} \qquad \rho = 1 .$$

Example 8.8. Let $f(z) = ze^z$. Here $N = 2$, $\rho = \gamma = T = 1$. Also

$$S_1 = \{\omega \mid (\mathrm{Re}\,\omega > -\tfrac{1}{2}) \cap (|\omega + \tfrac{2}{3}| \geqslant \tfrac{2}{3})\}$$

$$\sigma_1 = \{r \mid r \geqslant 0\}, \qquad m_\ell(\sigma_1) = \infty .$$

9. Solutions of Linear Differential Equations. We consider solutions w(z) of the linear differential equation

(9.1) $$L_n(w) = w^{(n)} + a_{n-1}(z)w^{(n-1)} + \cdots + a_0(z)w = 0 ,$$

where $a_j(z)$, $j = 0,1,\ldots,n-1$ are polynomials. It is known that every solution w is a polynomial or a transcendental entire function [70, Ch. 5]. More generally, if we consider the equation

(9.2) $$L_n^*(w) = P_0(z)w^{(n)} + P_1(z)w^{(n-1)} + \cdots + P_n(z)w = g(z) ,$$

with polynomial coefficients P_j and g also a polynomial, then every transcendental entire solution w satisfying (9.2) is of finite rational order and of perfectly regular growth, that is, there exist a positive number B and a rational number $\rho \geqslant 1/n$ such that

$$\lim_{r\to\infty} \frac{\log M(r)}{r^\rho} = B .$$

See [68, pp. 106–108]. If, in (9.2), the coefficients P_j $(j = 0,1,\ldots,n)$ are polynomials of degree not exceeding s and P_0 of degree s, and g a polynomial then every transcendental entire solution w of (9.2) is of order $\rho \leqslant 1$. If further, the coefficients of $w^{(n)}$ and w are both of deg s then $\rho = 1$ [70, Ch. 5].

We now consider theorems on solutions having b.i. and b.v.d.

Theorem 9.1 (Shah [53]). *Every solution* w *of the equation* $L_n(w) = 0$, *where* a_j, $j = 0,1,\ldots,n-1$ *are all constants, is a polynomial or a transcendental entire function, and has b.i. and b.v.d. Furthermore, the index*

$N = N_w \leq p$ *where* p *is any integer* $\geq n-1$ *and*

$$\frac{|a_{n-1}|}{p+1} + \frac{|a_{n-2}|}{p(p+1)} + \cdots + \frac{|a_0|}{(p-(n-2))\cdots(p+1)} \leq 1 \ .$$

The next theorem relaxes the hypothesis but the method does not yield a bound on the index.

Theorem 9.2 (Fricke and Shah [23]). *Let* $f(z)$ *be a transcendental entire function and suppose that it satisfies the equation*

$$(9.3) \qquad L_n^*(w) = g(z) \ ,$$

where $P_j(z)$, $j = 0,1,\ldots,n$ *are polynomials and* $P_0(z)$ *is of degree not less than that of any* $P_j(z)$, *and* $g(z)$ *is an entire function of b.i. Then the solution* $w = f(z)$ *is of b.i.*

Corollary 9.3. *If* $F(z)$ *is an entire function such that* $F'(z) = f(z)$ *satisfies the equation* (9.3) *then* $F(z)$ *is of b.v.d.*

Proof of Theorem 9.2. Since g is of b.i., by Theorem 4.1 there exist a positive integer N and a positive number c such that

$$|g^{(N+1)}(z)| \leq c \max_{0 \leq j \leq N} \{|g^{(j)}(z)|\}$$

for all $z \in \mathbf{C}$. Thus

$$\begin{aligned} |g^{(N+1)}(z)| &= \Big|\sum_{j=0}^{n} \frac{d^{N+1}}{dz^{N+1}} P_j(z) f^{(n-j)}(z)\Big| \\ &= \Big|\sum_{j=0}^{n} \sum_{k=0}^{N+1} \binom{N+1}{k} P_j^{(k)}(z) f^{(n-j+N+1-k)}(z)\Big| \\ &\leq c \max_{0\leq j\leq N} \{|g^{(j)}(z)|\} \\ &= c \max_{0\leq j\leq N} \left\{\Big|\sum_{k=0}^{n} \sum_{t=0}^{j} \binom{j}{t} P_k^{(t)}(z) f^{(n-k+j-t)}(z)\Big|\right\} \ . \end{aligned}$$

Hence

$$\begin{aligned} |P_0(z) f^{(n+N+1)}(z)| \leq \Big| &\sum_{j=1}^{n} \sum_{k=0}^{N+1} \binom{N+1}{k} P_j^{(k)}(z) f^{(n-j+N-1-k)}(z) \\ &+ \sum_{k=1}^{N+1} \binom{N+1}{k} P_0^{(k)}(z) f^{(n+N+1-k)}(z)\Big| \\ &+ c \max_{0\leq j\leq N} \left\{\Big|\sum_{k=0}^{n} \sum_{t=0}^{j} \binom{j}{i} P_k^{(t)}(z) f^{(n-k+j-t)}(z)\Big|\right\} \\ \leq\ &(c+1)(n+1)(N+2)(N+1)! \max_{0\leq j\leq n} \{|P_j^{(k)}(z)|\} \max_{0\leq j\leq N+n} \{|f^{(j)}(z)|\} \end{aligned}$$

Now since $\deg P_0 \geq \deg P_j$, $j = 0,1,\ldots,n$, there exist $M > 0$ and $R > 0$ such that $M|P_0(z)| \geq |P_j^{(k)}(z)|$ for all $j = 0,1,\ldots,n$ and $k = 0,1,\ldots,N+1$ and for all $|z| \geq R$. Thus

$$|f^{(n+N+1)}(z)| \leqslant (c+1)(n+1)(N+2)(N+1)!\,M \max_{0\leqslant j\leqslant N+n} \{|f^{(j)}(z)|\}$$

for $|z| \geqslant R$. Hence for $S = (c+1)(n+1)(N+2)(N+1)!\,M$ and $T = N+n$

$$|f^{(T+1)}(z)| \leqslant S \max_{0\leqslant j\leqslant T} \{|f^{(j)}(z)|\} \qquad \text{for } |z| \geqslant R .$$

This inequality holds for $|z| < R$ (cf. [54, pp. 132–133]) if we replace S by a suitable constant. Therefore, by Theorem 4.1, f is of b.i.

Proof of Corollary 9.3. Since f is of b.i., by Theorem 3.9 (see also [17]) F is of b.v.d.

Remarks. (i) The conclusions in Theorem 9.2 and Corollary 9.3 obviously hold if f is a polynomial.
(ii) Theorem 9.2 is applicable to entire solutions only. Consider, for instance,

$$z^2w'' - zw' - 8w = 0 .$$

Here the coefficients are polynomials and $\deg P_0 = 2 \geqslant \deg P_j$; but every solution is not entire. In fact, $w(z) = z^4$ is entire but $w(z) = z^{-2}$ is not.
(iii) If all the coefficients P_j are constants and g a polynomial in (9.3), then every solution is entire and of b.i. (Theorem 9.1). The following example shows that every solution of (9.3) may be entire and of b.i. but the coefficients P_j may not be constants.

Example 9.4. The equation $z^2w'' - 2zw' + 2w = 0$ has two solutions, $w_1(z) = d_1z$, $w_2(z) = d_2z + d_3z^2$, both polynomials and of b.i.

However, if we consider the equation (9.1), $L_n(w) = 0$, then we have the following:

Theorem 9.5 (Wittich [71]). *Suppose that every solution of the equation* (9.1), $L_n(w) = 0$, *with all* a_j *entire, has b.v.d. Then the coefficients* $a_0, a_1, \ldots, a_{n-1}$ *are necessarily constants.*

More generally, suppose that there exists an $R > 0$ *such that for every solution* $w(z)$ *of the equation, there exists at least one* $c \neq 0$, $c \in \mathbf{C}$, *for which*

$$\sup_a n\left(a, R, \frac{1}{w-c}\right) < \infty ,$$

then the coefficients $a_0, \ldots, a_{n-1}$ *are constants.*

Theorem 9.6 (Frank and Mues [11]). *Every solution of the equation* $L_n(w) = 0$, *with all* a_j *polynomials, is of b.i. if and only if the equation has constant coefficients.*

The next theorem gives a bound for $T(w)$.

Theorem 9.7 (Lee and Shah [39]). *Let* $w(z)$ *be an entire solution of the equation* $L_n^*(w) = 0$. *Let* $P_j(z) = a_jz^s + \cdots$, $j = 0,1,\ldots,n$ *and* $a_0 \neq 0$. *Let* q *be the least nonnegative integer such that*

$$(k+q)!\,|a_0| > \sum_{j=1}^{k} (k+q-j)!\,|a_j| .$$

Then

$$T(w) = \limsup_{r\to\infty} \frac{\log M(r,w)}{r} \leqslant k+q \; .$$

The following example shows that every entire solution of the equation $L_n^*(w) = 0$ may be of b.i. but deg P_0 may be less than $\max_{j>0}\{\deg P_j\}$.

Example 9.8. $zw'' + (z^2 - z - \frac{1}{2})w' - (z^2 - \frac{1}{2})w = 0$. Then one solution is $w_1(z) = e^z$ which is of b.i. and of b.v.d. and the second solution is not entire.

10. Functions having b.i. and b.v.d.

(a) The Bessel function $J_k(z)$, k integer or zero, is entire and satisfies the differential equation

$$L_2^*(w) = z^2w'' + zw' + (z^2 - k^2)w = 0$$

and $\deg P_0 \geqslant \deg P_j$ $(j = 0,1,2)$.

Hence $J_k(z)$ has b.i. The same argument shows that $J_k'(z)$ has b.i. Furthermore, the inequalities satisfied by the Bessel function and the derivatives enable one to find bounds on the index.

Theorem 10.1 (Lee and Shah [40]). *Let* k *be an integer or zero. The Bessel function* $J_k(z)$ *has b.i. and b.v.d. The index* N *of* $J_k(z)$ *when* k *is zero or a positive integer satisfies* $k \leqslant N \leqslant \max(1, 2k-1)$.

Theorem 10.2 (Marić and Shah [45]). *Let* N *denote the index of the entire function* $z^{-k}J_k(z)$ *where* k *is any nonnegative number. Then*

(i) $N = 1$ if $0 \leqslant k \leqslant 0.21$,

(ii) $1 \leqslant N \leqslant 3$ if $0.21 \leqslant k \leqslant 2.31$,

(iii) $1 \leqslant N \leqslant \max\left(4, \frac{2k}{1.1}\right)$ otherwise.

This theorem has been improved by Tang [67]. Note that $w(z) = z^{-k}J_k(z)$ satisfies the equation

$$zw'' + (1 + 2k)w' + zw = 0$$

and $w'(z)$ also satisfies an equation satisfying the conditions of Theorem 9.2. Hence the entire function

$$z^{-k}J_k(z) = (\tfrac{1}{2})^k \sum_{p=0}^{\infty} \frac{(-\frac{1}{4}z^2)^p}{p!\,\Gamma(k+p+1)} \; ,$$

where k is any nonnegative number has b.i. and b.v.d.

(b) Any entire solution of the equation (9.2), $L_n^*(w) = 0$, with polynomial coefficients and $\deg P_0 \geqslant \deg P_j$ $(j = 0,1,2,\ldots,n)$ has b.i. and b.v.d.

(c) Consider the functions ψ and Ψ defined by

(10.1) $$\psi(z) = \int_a^b e^{zt}\varphi(t)\,dt \; ,$$

(10.2) $$\Psi(z) = \sum_{k=1}^{n} P_k(z)\exp(a_k z) + \psi(z) \; ,$$

where the P_k's are polynomials, $a_k \in \mathbf{C}$, a_k's all distinct, $-\infty < a < b < \infty$ and $\varphi \in L[a,b]$. It is easily verified that ψ and Ψ are both entire functions of exponential type [1, pp. 8, 108].

Theorem 10.3 (Shah [59]). *Let* $\varphi(t)$ *be a complex-valued function absolutely continuous on* $[a,b]$ *and such that* $\varphi(a) \neq 0$, $\varphi(b) \neq 0$. *Then the function* ψ *defined by* (10.1) *is an entire function of b.i. Furthermore, if* $ab \neq 0$, ψ *and all successive derivatives* $\psi^{(k)}$ *have b.i. and b.v.d.*

It is possible to replace the conditions in Theorem 10.3 by somewhat less restrictive conditions ([59]). The proof utilizes the information on the zeros of ψ.

Theorem 10.4 (Shah [59]). *Let* φ *satisfy the conditions of Theorem 10.3 and let the* P's *be polynomials and* $a_k \neq a$ *or* b *for* $k = 1,2,\ldots,n$. *Then the function* Ψ *defined by* (10.2) *is entire and has b.i. Furthermore, if* $ab \neq 0$, *then* Ψ *and all successive derivatives* $\Psi^{(k)}$ *have b.i. and b.v.d.*

Theorem 10.5 (Shah [59]). *Let* $a_j, \alpha_j \in \mathbf{C}$. *Suppose that* $\varphi^{(k)} \in L[a,b]$ *and* φ *satisfies the equation*

$$\sum_{j=0}^{k} P_j^*(t)\varphi^{(j)}(t) = \sum_{j=1}^{m} Q_j(t)\exp(\alpha_j t)$$

where P_j^* *and* Q_j *are polynomials, the* α*'s are all distinct and* $\deg P_k^* \geqslant \deg P_j^*$. *Then* Ψ *defined by* (10.2) *is entire and* Ψ *and all successive derivatives* $\Psi^{(p)}$ *have b.i. and b.v.d.*

(d) In the next group of theorems we are able to find a bound on the index.

Theorem 10.6 (Ekblaw [5]). *Let* $f_1,\ldots,f_k$ *be entire functions such that for* $1 \leqslant j \leqslant k$, *there exists an integer* n_j *and a complex number* a_j *such that* $f_j^{(n_j)} = a_j f$. *Let* p_j $(j = 1,2,\ldots,k)$ *be a polynomial. If* $g(z) = \Sigma_{j=1}^{k} p_j(z)f_j(z)$ *then* g *has b.i. and b.v.d.*

Theorem 10.7 (Ekblaw [5]). *Let* f *be entire and* $n \geqslant 1$ *be the least integer such that* $f^{(n)} = af$ $(a \in \mathbf{C})$. *Let* k *be a nonnegative integer such that* $|a| \leqslant (k+1)\cdots(k+n)$. *Then* f *is of index* $\leqslant k+n-1$.

Example 10.8. Let $f(z) = \cos 2z$. Then $n = 2$, $a = 4$, $k = 1$, $N = 2$. This shows that the bound for N is sharp.

Theorem 10.9 (Ekblaw [5]). *Let* f *be entire and* $f^{(n)} = af$. *Let* k *be a nonnegative integer such that* $|a| \leqslant (k+1)\cdots(k+n)$, *and* p *a polynomial of degree* m. *Then* $(f+p)$ *is of index* $\leqslant k+n+m$.

Theorem 10.10 (Ekblaw [5]). *Let* f *be entire,* $|a| \leqslant 1$ *and* $zf^{(n)}(z) = af(z)$. *If* $f^{(k)}(0) = 0$ *for* $k = 2,3,\ldots,(n-1)$, *then* f *is of index* $\leqslant \max(9,n)$.

11. Functions of b.i. as a metric space.

In [33] Iyer considers the class Γ of all entire functions (including polynomials and the constant function that is identically zero) topologized in the following manner. Let $f(z) = \Sigma_{n=0}^{\infty} a_n z^n$ and $g(z) = \Sigma_{n=0}^{\infty} b_n z^n$ be two entire functions. Let

$$d(f,g) = \sup\{|a_0 - b_0|, |a_n - b_n|^{1/n}, n = 1,2,\ldots\}.$$

Then d is a metric on the class Γ, and the space (Γ,d), of entire functions with the topology generated by d, is a complete metric space. Now consider a subclass of Γ, namely functions of b.i. Let

$$(B_n,d) = \{f \in (\Gamma,d) \mid f \text{ is of index} \leqslant n\}$$

$$(B,d) = \cup_{n=1}^{\infty} B_n .$$

Theorem 11.1 (Ekblaw [6]). *The space* (B,d) *of entire functions of bounded index is a first category space.*

For further results of this type see [5, 64].

12. Functions of several variables. In the complex n-space C^n we define functions of bounded index by requiring, as in Definition 1.1, that the Taylor coefficients satisfy inequalities analogous to (1.3). See [37]. An alternate way of defining functions of b.i. has been suggested by Sisarcick [64]. Consider, for instance, an entire function f(z,w) in two variables. Let E^2 denote the space of entire functions f(z,w) such that $f \in E^2$ if and only if there exists a positive integer N and a constant $C > 0$ such that for all (z,w) and all j,k such that $j+k = p$, $p = 0,1,\ldots,$

$$\sum_{j+k=0}^{N} \frac{\left|\dfrac{\partial^{j+k} f(z,w)}{\partial z^j \partial w^k}\right|}{j!k!} \geqslant C \sum_{j+k=N+1}^{\infty} \frac{\left|\dfrac{\partial^{j+k} f(z,w)}{\partial z^j \partial w^k}\right|}{j!k!} .$$

Compare with Theorem 4.2.

Let Γ^2 denote the space of entire functions of two variables. Let $f(z,w) = \Sigma_{m,n=0}^{\infty} a_{m,n} z^m w^n$ and $g(z,w) = \Sigma_{m,n=0}^{\infty} b_{m,n} z^m w^n$. Let

$$e(f,g) = \sup \{|a_{00} - b_{00}|, |a_{m,n} - b_{m,n}|^{1/(m+n)}, m+n \geqslant 1\} .$$

Then e is a metric on Γ^2 and the space (Γ^2,e) is a metric space.

Theorem 12.1 (Sisarcick [64]). *The space* (E^2,e) *is a first category space.*

Finally, for vector-valued entire functions of b.i. we refer the reader to [5, 32] and for functions of b.i. over non-Archimedean field to [66], and conclude with a few open problems.

1. Consider an entire solution of the equation $L_n^*(w) = 0$ with polynomial coefficients P_j and $\deg P_o \geqslant \deg P_j$ $(j = 0,1,\ldots,n)$. Find a bound for the index. (See Theorems 9.2 and 10.1.)

2. Does the b.i. imply that $T = \gamma$? (See Section 2.)

3. Consider the function ψ defined in (10.1) for b.i. and b.v.d., when $\varphi(a) = 0$, $\varphi(b) \neq 0$, and also when $a = 0$, $b = \infty$. Extend Theorems 10.3 and 10.4.

4. Let φ be any entire function of exponential type. Does there exist an entire function f of b.i. such that $\log M(r,\varphi) \sim \log M(r,f)$ as $r \to \infty$?

5. Let $k > 1$. Prove that the Mittag-Leffler function F(z,k), where k is not necessarily an integer, has b.i. (See Theorem 6.14. We assumed there that k is an integer.)

References

1. R. P. Boas, **Entire Functions**, Academic Press, New York, 1954.

2. Anil K. Bose, A note on entire functions of bounded index, *Proc. Amer. Math. Soc.* **21** (1969), 257–262.

3. M. L. Cartwright, **Integral Functions**, Camb. Univ. Press, 1962.

4. C. S. Davis, Entire functions of bounded absolute index, Doctoral dissertation, Catholic Univ. of America, Washington, D.C., 1973.

5. K. A. Ekblaw, The functions of bounded index as a subspace of a space of entire functions, Doctoral dissertation, Univ. of Kentucky, Lexington, 1970.

6. K. A. Ekblaw, The functions of bounded index as a subspace of a space of entire functions, *Pacific J. Math.* **37** (1971), 353–355.

7. K. A. Ekblaw, Polynomials of maximal index, *Aequationes math.* **10** (1974), 34–39.

8. Günter Frank, Picardsche Ausnahmewerte bei Lösungen linearer Differentialgleichungen, Dissertation, Universitat Karlsruhe (1969).

9. Günter Frank, Über den Index einer ganzen Funktion, *Arch. der Math.* **22** (1971), 175–180.

10. Günter Frank, Über ganze Funktionen, die einer linearen Differentialgleichung genügen, *manuscripta math.* **4** (1971), 225–253.

11. Günter Frank and Erwin Mues, Über den Index der Lösungen linearer Differentialgleichungen, *manuscripta math.* 5 (1971), 155–163.

12. Günter Frank and Erwin Mues, Über das Wachstum des Index ganzer Funktionen, *Math. Ann.* **195** (1972), 114–120.

13. Günter Frank and Erwin Mues, Über lokale Maximalbeträge und den Index ganzer Funktionen, *Arch. der Math.* **23** (1972), 269–277.

14. Günter Frank, Zur lokalen Werteverteilung der Lösungen linearer Differentialgleichungen, *manuscripta math.* **6** (1972), 381–404.

15. G. H. Fricke and R. E. Powell, Some product theorems on functions of bounded index, *Indian J. Pure Appl. Math.* **3** (1972), 795–800.

16. Gerd H. Fricke, A characterization of functions of bounded index, *Indian J. Math.* **14** (1972), 207–212.

17. Gerd H. Fricke, A note on multivalence of functions of bounded index, *Proc. Amer. Math. Soc.* **40** (1973), 140–142.

18. Gerd H. Fricke, Functions of bounded index and their logarithmic derivatives, *Math. Ann.* **206** (1973), 215–223.

19. G. H. Fricke, S. M. Shah and W. C. Sisarcick, A characterization of entire functions of exponential type and M-bounded index, *Indiana Univ. Math. J.* **23** (1973), 405–412.

20. Gerd H. Fricke, Entire functions having positive zeros, *Indian J. Pure and App. Math.* 5 (1974), 478–485.

21. Gerd H. Fricke, Entire functions of locally slow growth, *J. d'Analyse Math.* **28** (1975), 101–122.

22. Gerd H. Fricke, A characterization of functions having positive zeros, *Indian J. Math.*, to appear.

23. G. H. Fricke and S. M. Shah, Entire functions satisfying a linear differential equation, *Indag. Math.* **37** (1975), 39–41.

24. G. H. Fricke and R. E. Powell, Bounded index and summability methods, *J. Austral. Math. Soc. Ser. A* **21** (1976), 79–87.

25. Gerd H. Fricke, Entire functions with prescribed asymptotic behavior, *Rocky Mountain J. Math.* **6** (1976), 237–246.

26. A. W. Goodman, The valence of sums and products, *Canad. J. Math.* **20** (1968), 1173–1177.

27. Fred Gross, Entire functions of bounded index, *Proc. Amer. Math. Soc.* **18** (1967), 974–980.

28. Fred Gross, Entire functions of exponential type, *J. Res. Nat. Bur. Standards Section B* **74B** (1970), 55–59.

29. W. K. Hayman, **Research problems in Function Theory**, Athlone Press, London, 1967.

30. W. K. Hayman, Differentialgleichungen und lokale Wurzelen, *Meh. spolšn, Sredy rodstv. Probl. Analiz.*, 591–595 (Russian), (1972).

31. W. K. Hayman, Differential inequalities and local valency, *Pacific J. Math.* **44** (1973), 117–137.

32. L. F. Heath, Vector-valued entire functions of bounded index satisfying a differential equation, to appear.

33. V. G. Iyer, On the space of integral functions, I, *J. Indian Math. Soc.* **12** (1948), 13–30.

34. A. C. King, A class of entire functions of bounded index and radii of univalence of some functions of zero order, Doctoral dissertation, Univ. of Kentucky, Lexington, 1970.

35. A. C. King and S. M. Shah, Indices of Lindelöf functions and their derivatives, *Rocky Mountain J. Math.* **2** (1972), 579–594.

36. O. Knab, Wachstumsordnung und Index der Lösungen linearer Differentialgleichungen mit rationalen Koeffizienten, *manuscripta math.* **18** (1976), 299–316.

37. J. G. Krishna and S. M. Shah, Functions of bounded indices in one and several complex variables, *Macintyre Memorial Volume*, Athens, Ohio, 223–235, (1970).

38. T. V. Lakshminarasimhan, A note on entire functions of bounded index, *J. Indian Math. Soc.* **38** (1974), 43–49.

39. Boo-sang Lee and S. M. Shah, The type of an entire function satisfying a linear differential equation, *Arch. der Math. Basel* **20** (1969), 616–622.

40. Boo-sang Lee and S. M. Shah, An inequality involving the Bessel function and its derivatives, *J. Math. Anal. and Appl.* **30** (1970), 144–155.

41. Boo-sang Lee and S. M. Shah, Inequalities satisfied by entire functions and their derivatives, *Trans. Amer. Math. Soc.* **149** (1970), 109–117.

42. Boo-sang Lee and S. M. Shah, On the growth of entire functions of bounded index, *Indiana Univ. Math. J.* **20** (1970/71), 81–87.

43. B. Lepson, Differential equations of infinite order, hyperdirichlet series and entire functions of bounded index, *Proc. Sympos. Pure Math.* **vol. XI**, Amer. Math. Soc., Providence, R.I., (1968), 298–307.

44. J. S. J. MacDonnell, Some convergnece theorems for Dirichlet series whose coefficients are entire functions of bounded index, *Catholic Univ. Press,* Washington, D.C., 1957.

45. V. Maric and S. M. Shah, Entire functions defined by gap power series and satisfying a differential equation, *Tohoku Math. J.* (2) **21** (1969), 621–631.

46. J. Nikolaus, Über ganze Losungen linearer Differentialgleichungen, *Archiv d. Math.* **18** (1967), 618–626.

47. J. Nikolaus, Lineare Differentialgleichungen mit gegebener ganzer Lösung, *Math. Zeit* **103** (1968), 30–36.

48. G. Polya and G. Szegö, **Problems and Theorems in Analysis, Vol. II**, Springer-Verlag, New York, 1976.

49. Walter Pugh, Sums of functions of bounded index, *Proc. Amer. Math. Soc.* **22** (1969), 319–323.

50. W. J. Pugh and S. M. Shah, On the growth of entire functions of bounded index, *Pacific J. Math.* **83** (1970), 191–201.

51. Q. I. Rahman and L. Stankiewicz, Differential inequalities and local valency, *Pacific J. of Math.* **LIV**, No. 2, (1974), 165–182.

52. G. Sansone and J. Gerretsen, **Lectures on the Theory of Functions of a Complex Variable, vol. I**, P. Noordhoff, Gronigen, 1960.

53. S. M. Shah, Entire functions of bounded index, *Proc. Amer. Math. Soc.* **19** (1968), 1017–1022.

54. S. M. Shah, Entire functions satisfying a linear differential equation, *J. Math. Mech.* **18** (1968/69), 131–136.

55. S. M. Shah, Entire functions of unbounded index and having simple zeros, *Math. Zeit.* **118** (1970), 193–196.

56. S. M. Shah, On entire functions of bounded index whose derivatives are of unbounded index, *J. London Math. Soc.* (2) **4** (1971), 127–139.

57. S. M. Shah, Analytic functions with univalent derivatives and entire functions of exponential type, *Bull. Amer. Math. Soc.* **78** (1972), 154–171.

58. S. M. Shah, Functions of bounded index and M-bounded index, *J. of Math. and Physical Sciences* **7** (1973), Silver Jubilee Issue, S81–S86.

59. S. M. Shah, Entire functions whose Fourier transforms vanish outside a finite interval, *J. Math. Anal. and Applications* **53** (1976), 174–185.

60. S. M. Shah and S. N. Shah, A new class of functions of bounded index, *Trans. Amer. Math. Soc.* **173** (1972), 363–377.

61. S. M. Shah and W. C. Sisarcick, On entire functions of exponential type, *J. Res. Nat. Bur. Standards,* Sect. B **75B** (1971), 141–147.

62. S.M. Shah and S.Y. Trimble, Entire functions with univalent derivatives, *J. Math. Anal. and Appl.* **33**(1971), 220–229.

63. Shantilal N. Shah, Functions of exponential type are differences of functions of bounded index, *Canad.J. Math.,* to appear.

64. W.C. Sisarcick, Variations of the definition of a function of bounded index, Doctoral dissertation, Univ. of Kentucky, Lexington, 1971;

65. W.C. Sisarcick, Metric spaces of entire functions, *Indian J. of Pure and Applied Math.,* to appear.

66. V. Sreenivasulu, Entire functions of bounded index over non-Archimedean field, *Annales Polonici Math.* **28**(1973), 163–173.

67. Hsiung Tang, Entire functions related to a Bessel function, *Tamkang J. of Math.,*Taiwan, to appear.

68. Georges Valiron, **Lectures on the General Theory of Integral Functions**, Chelsea, New York, 1949.

69. J.M. Whittaker, The lower order of integral functions, *J. London Math. Soc.* **8**(1933), 20–27.

70. Hans Wittich, **Neuere Untersuchungen über Eindeutige Analytische Funktionen**, Springer-Verlag, Berlin, 1955.

71. Hans Wittich, Zur Kennzeichung linearer Differentialgleichungen mit konstanten Koeffizienten, *Festband Zum,* **70** Geburtstag von R. Nevanlinna, Springer, Berlin, 128–134, 1966.

Added in proof:

(i) Consider the equation $L_n(w) = 0$, defined in Section 9, where $a_j(z)$ are all entire functions. Tijdeman has shown that these coefficients a_j are polynomials if and only if there exist fixed numbers p and q such that every solution $w(z)$ of the equation is p-valent in any disk

$$\{z: |z - z_0| < 1/(1 + r^q)\}, \text{ where } r = |z_0|.$$

See [29, p. 17; 72, p. 147; 73].

(ii) Turan asked whether it was sufficient to make the assumption (3.5) not for all values w but only for 3 such values, to assure that $f(z)$ is b.v.d. A negative answer was provided by Gol'dberg. Another such example is given by Weierstrass sigma function which is an entire function of order 2 and so cannot be a b.v.d. function. See [29, p. 17; 72, p. 147].

References

72. J.C. Clunie and W.K. Hayman, Symposium on Complex Analysis Canterbury, Lond. Math. Soc. Lecture Notes, 1974.

73. R. Tijdeman, On the distribution of values of certain functions, Doctoral dissertation, Univ. of Amsterdam, 1969.

University of Kentucky
Lexington, Kentucky 40506

STARLIKENESS, CONVEXITY AND OTHER GEOMETRIC PROPERTIES OF HOLOMORPHIC MAPS IN HIGHER DIMENSIONS

T. J. Suffridge

1. Introduction. In the study of the functions $f(z) = \Sigma_{k=1}^{\infty} a_k z^k$ that are analytic and univalent in the unit disk $\{|z| < 1\}$, certain subclasses in which $f(|z| < 1)$ has some simple geometric property arise rather naturally. For example, if f is analytic in $\{|z| < 1\}$, $f'(0) \neq 0$ and $\mathrm{Re}(zf'(z)/f(z)) > 0$ when $|z| < 1$ then f is univalent and $f(|z| < 1)$ is starlike with respect to the origin. Thus, a rather simple analytic condition $(\mathrm{Re}(zf'(z)/f(z)) > 0)$ implies univalence of f as well as the geometric property of starlikeness with respect to the origin of the image region. Other geometric properties of the image region that are implied by simple analytic conditions on the mapping function are convexity, close-to-convexity and spirallikeness.

In this paper, we consider the extension of these ideas to higher dimensions. The results will usually be given for functions that are holomorphic and univalent in the unit ball of a Banach space. In some cases, compactness of the closed unit ball is required and in these cases, the additional assumption of finite dimensionality is made. The results are generally given in a coordinate free form. However, in some instances, the finite dimensional case with the usual coordinates is of special interest and the particular form of the result for this case is given. The literature on this subject is rather limited (see [4], [9], [10], [11], [15], and [16]). We will survey the known results and give a number of examples as well as giving some new results on close-to-convexity and spirallikeness.

2. Notation and Preliminary Results. The symbols X and Y will be used to denote Banach spaces while $B_r = \{x \in X: \|x\| < r\}$ and $S_r = \{x \in X: \|x\| = r\}$. Further, we set $B = B_1$ and $\overline{B}_r = B_r \cup S_r = \{x \in X: \|x\| \leq r\}$.

If U is an open subset of X and $f: U \to Y$ then f is holomorphic in U if given $x \in U$, there is a bounded linear map $Df(x): X \to Y$ such that

$$\lim_{h\to 0} \frac{\|f(x+h) - f(x) - Df(x)(h)\|}{\|h\|} = 0 .$$

The linear map $Df(x)$ is called the Fréchet derivative of f at x. In case $X = Y = \mathbf{C}$ (the complex plane), the Fréchet derivative of an analytic function f at a point $z \in U$ is the linear map which multiplies by $f'(z)$. In case $X = \mathbf{C}^n$ and $Y = \mathbf{C}^k$, the Fréchet derivative of a holomorphic map f at a point $(z_1, z_2, \ldots, z_n) \in U$ is the linear map which has the Jacobian matrix of the map f as its matrix. The reader is referred to [6, Chapters 3 & 26] for further elementary properties of holomorphic maps on Banach spaces. We mention here that if f is holomorphic in U and $x \in U$ then for $n = 1,2, \ldots,$ there is a bounded symmetric n-linear map $D^n f(x): X \times X \times \cdots \times X \to Y$ such that $f(y) = \Sigma_{n=0}^{\infty} \frac{1}{n!} D^n f(x)((y-x)^n)$ for all y in some neighborhood of x. Here $D^0 f(x)((y-x)^0) = f(x)$ and for $n \geq 1$, $D^n f(x)((y-x)^n) = D^n f(x)(y-x, y-x, \ldots, y-x)$.

This research was supported in part by the National Science Foundation under grant number MPS 75-06971.

A function $f: U \to Y$ is biholomorphic if the inverse function f^{-1} exists and is holomorphic on an open set $V \subset Y$ and $f^{-1}(V) = U$. A function f is locally biholomorphic if given $y \in f(U)$ there is a neighborhood V of y such that f^{-1} exists and is holomorphic in V. It is a standard result that $Df(x)$ has a bounded inverse for each $x \in U$ if and only if f is locally biholomorphic. In finite dimensional spaces, if $Y = X = \mathbf{C}^n$, the condition that $Df(x)$ have a bounded inverse is just the condition that the determinant of the Jacobian be non-zero.

Example 1. Note that it is possible for a holomorphic function to be univalent in B and have the zero linear map for its derivative at at least one point. For example, if $f: \{|z| < 1\} \to \mathbf{C}^2$ is given by $f(z) = (z^2, z^3)$, then $f(z) = f(w) \Rightarrow z = z^3/z^2 = w^3/w^2 = w$ if $zw \neq 0$ while $z = 0$ if and only if $w = 0$. Thus f is univalent. However, $Df(0)$ is the zero linear map from $\mathbf{C}$ to $\mathbf{C}^2$.

In fact, if $X = \ell^2 = Y$ (here ℓ^2 is the Hilbert space of square summable sequences) and $f: B \to X$ is defined by $f(z_1, z_2, \ldots) = (z_1^2, z_1^3, z_2^2, z_2^3, \ldots)$ then f is univalent and holomorphic on the unit ball of X with values in X and $Df(0)$ is the zero linear map of X into itself.

This example shows that the result in the plane—if f is analytic in a neighborhood of z and $f'(z) = 0$ then f is not univalent in any neighborhood of z —is not true when the dimension of the range space is greater than that of the domain space or when the domain space is infinite dimensional. However, it is true that if U is an open set in $\mathbf{C}^n$ and $f: U \to \mathbf{C}^n$ then f is locally biholomorphic if and only if the Jacobian determinant is non-zero [3, pp. 179–182].

There is more than one version of "Schwarz lemma" in normed linear spaces (see [5]). We require the following version which is well-known. We include a proof since it is short and involves a commonly used technique.

Theorem A. (Schwarz lemma) *If* $f: B \to Y$ *is holomorphic,* $\|f(x)\| \leq 1$ *when* $x \in B$ *and* $f(0) = 0$ *then* $\|f(x)\| \leq \|x\|$ *for all* $x \in B$.

Proof. Let x be a fixed element of B and let $\ell: Y \to \mathbf{C}$ be a continuous linear functional such that $\ell(f(x)) = \|f(x)\|$ and $|\ell(y)| \leq \|y\|$ for all $y \in Y$ (such a functional exists by the Hahn-Banach theorem). Then the function g defined by $g(\alpha) = \ell(f((\alpha/\|x\|)x))$ is complex valued and analytic in the unit disk, $g(0) = 0$ and $|g(\alpha)| = |\ell(f((\alpha/\|x\|)x))| \leq \|f((\alpha/\|x\|)x)\| \leq 1$. Therefore, by Schwarz's lemma in the plane, $|g(\alpha)| \leq |\alpha|$. Setting $\alpha = \|x\|$, we see that $\|f(x)\| = |\ell(f(x))| = |g(\|x\|)| \leq \|x\|$ and the proof is complete.

Since the results that we wish to extend generally involve a function with positive real part, we wish to extend this concept to higher dimensions. The extension is made by the use of linear functionals. Let X^* be the dual of X and given $x \in X, x \neq 0$, define $T(x) = \{x^* \in X^*: x^*(x) = \|x\| \text{ and } \|x^*\| = 1\}$. Note that $T(x)$ is non-empty. Also, observe that given $x \neq 0, x \in X, x^* \in T(x)$, the hyperplane $H_1 = \{y \in X: \operatorname{Re} x^*(y) = \|x\|\}$ is a supporting hyperplane for the ball $B_{\|x\|}$ at the point x for $x \in H_1$ by definition of $T(x)$ while $y \in H_1 \Rightarrow \|y\| \geq |x^*(y)| = \|x\|$. The half-space $H = \{y \in X: \operatorname{Re} x^*(y) > 0\}$ is the set of all points "on one side" of the hyperplane through the origin and "parallel" to H_1.

Let N denote the set of all holomorphic functions $w: B \to X$ such that $w(0) = 0$ and $\operatorname{Re} x^*(w(x)) > 0$

when $x \in B$, $x \neq 0$ and $x^* \in T(x)$. Let $M = \{w \in N: Dw(0) = I\}$ (here I is the identity map on X). Thus M is the set of functions in N that have series developments of the form $w(x) = x + \Sigma_{n=2}^{\infty} \frac{1}{n!} D^n w(0)(x^n)$. This notation is the notation of Gurganus [4].

It is helpful to one's intuition to characterize the sets $T(x)$, N and M in case $X = \mathbf{C}$. Given $u \in \mathbf{C}$, $u \neq 0$, we wish to determine $T(u)$. Any complex linear functional ℓ on $\mathbf{C}$ has the form $\ell(z) = \alpha z$ for some α. Since we require $\ell(u) = |u|$, we conclude $\alpha u = |u|$ and $\alpha = |u|/u$. Thus, as expected, $u^* \in T(u)$ is unique and has the form $u^*(z) = (|u|/u)z$. The equation $\mathrm{Re}\,((|u|/u)z) = |u|$ (u fixed, $u \neq 0$) is the equation of a line tangent to the circle $\{z: |z| = |u|\}$ at the point $z = u$. Further, $w \in N \Rightarrow w$ is analytic in $\{|z| < 1\}$, $w(0) = 0$, and $\mathrm{Re}((|z|/z)w(z)) > 0$ when $0 < |z| < 1$. Since $w(0) = 0$, $w(z)/z$ is analytic at 0 (if properly defined there) and it is clear that $N = \{zp(z)$: p is analytic in the unit disk and $\mathrm{Re}\, p(z) > 0\}$. Further, $M = \{zp(z) \in N: p(0) = 1\}$.

The class N can be described geometrically as being the class of holomorphic functions $w: B \to X$ which fix the origin and map the point $x \neq 0$ into the intersection of all half-spaces determined as follows. Let H_1 be a supporting hyperplane (of real codimension 1) to the ball $B_{\|x\|}$ at the point x. Let H be a parallel hyperplane through the origin. Then x and $w(x)$ lie on the same side of H.

It is convenient to set $N_0 = \{w$: w is holomorphic in B with values in X, $w(0) = 0$ and if $x \in B$, $x \neq 0$, $x^* \in T(x)$ then $\mathrm{Re}\, x^*(w(x)) \geqslant 0\}$ (i.e. N_0 differs from N only in the fact that one allows $\mathrm{Re}\, x^*(w(x)) \geqslant 0$ rather than $\mathrm{Re}\, x^*(w(x)) > 0$). Thus $N_0 \supset N$. In case $X = \mathbf{C}$, by the minimum principle we conclude $w \in N_0 - N$ if and only if w has the form $w(z) = itz$ for some real t (i.e. $w(z)/z$ is a constant with zero real part). Example 2 below shows that $N_0 - N$ need not be so trivial if $X \neq \mathbf{C}$. The reader can readily verify (for a proof, see [16, Lemma 3]) that if $w \in N_0$ then $w_\alpha \in N_0$ when $|\alpha| < 1$ where $w_\alpha(x) = \frac{1}{\alpha} w(\alpha x)$ ($= Dw(0)(x)$ when $\alpha = 0$). The proof consists of using the fact that if $x^* \in T(x)$ then $x^*_\alpha \in T(\alpha x)$ where x^*_α is defined by $x^*_\alpha(y) = x^*((|\alpha|/\alpha)y)$ for all $y \in X$. The minimum principle for harmonic functions of one variable then shows that if $w \in N_0$ then $w \in N$ if and only if $Dw(0)(\cdot) \in N$ (again, see [16, Lemma 3]). Thus, to determine whether a function in N_0 is actually a function of N, one need only consider the linear terms $Dw(0)(x)$.

Example 2. If $X = \mathbf{C}^2$ with Euclidean norm: $\|(z_1,z_2)\|^2 = |z_1|^2 + |z_2|^2$, then given $u = (u_1,u_2) \in X$, $(u_1,u_2) \neq (0,0)$, the set $T(u_1,u_2)$ consists of exactly one element ℓ_u given by $\ell_u(z_1,z_2) = (\bar{u}_1 z_1 + \bar{u}_2 z_2)/\|(u_1,u_2)\|$. The function $w: B \to X$ given by $w(z_1,z_2) = (-z_2,z_1)$ is an element of $N_0 - N$ since $\mathrm{Re}\, \ell_z(w(z)) = \mathrm{Re}\,(-\bar{z}_1 z_2 + \bar{z}_2 z_1)/\|(z_1,z_2)\| = 0$, $z = (z_1,z_2) \neq (0,0)$.

Two lemmas which will be useful in our further considerations are given in [16, Theorems A and B]. These are generalizations to higher dimensions of results of Robertson [13, Theorems A and B]. A proof will be given of the first of these since it is rather short and easy while an outline of the method of proof of the second will be given.

Lemma 1. *Let* $v: B \times I \to B$ *be holomorphic in* B *for each* $t \in I = [0,1]$ (*i.e.* $v(\cdot,t)$ *is holomorphic in* B), $v(0,t) = 0$ *and* $v(x,0) = x$. *If* $\lim_{t \to 0+} (x - v(x,t))/t = w(x)$ *exists and is holomorphic in* B *then* $w \in N_0$.

Proof. Assume $x \in B$, $x \neq 0$ and $x^* \in T(x)$. By Schwarz's lemma, $\|v(x,t)\| \leq \|x\|$ so that $\mathrm{Re}\, x^*((x - v(x,t))/t) = \mathrm{Re}\,(\|x\| - x^*(v(x,t))/t \geq (\|x\| - \|v(x,t)\|)/t \geq 0$ and the desired result follows by continuity of x^*.

Non-trivial examples in which $w \in N_0 - N$ can readily be found. For example, if $X = C^2$ with Euclidean norm, one can choose $v(x,t)$ to be an appropriate unitary map (so $\|v(x,t)\| = \|x\|$). For example (see [16, p. 577]) if for fixed t, $v(x,t)$ is the restriction to B of the linear map associated with the matrix

$$\begin{pmatrix} \sqrt{1-t^2} & t \\ -t & \sqrt{1-t^2} \end{pmatrix}$$

then $w(x) = (-x_2, x_1)$ which is the function given in Example 2.

Lemma 2. *Let* $f\colon B \to Y$ *be a biholomorphic map of* B *onto an open set* $f(B) \subset Y$ *and let* $f(0) = 0$. *Assume* $F\colon B \times I \to Y$ *is holomorphic in* B *for each fixed* $t \in I$, $F(x,0) = f(x)$, $F(0,t) = 0$ *and suppose* $F(B,t) \subset f(B)$ *for each fixed* $t \in I$. *Further, suppose*

$$\lim_{t \to 0+} \frac{F(x,0) - F(x,t)}{t} = G(x)$$

exists and is holomorphic. Then $G(x) = DF(x)(w(x))$ *where* $w \in N_0$.

The proof consists of showing that the function $v(x,t) = f^{-1}(F(x,t))$ satisfies the hypotheses of Lemma 1. We have $F(x,t) = f(v(x,t)) = f(x) + Df(x)(v(x,t) - x) + R(v(x,t),x)$ where R has the property $\|R(y,x)\|/\|y - x\| \to 0$ as $\|y - x\| \to 0$. Thus,

$$\frac{F(x,0) - F(x,t)}{t} = Df(x)\left(\frac{x - v(x,t)}{t}\right) - \frac{R(v(x,t),x)}{t}.$$

One shows that

$$\lim_{t \to 0+} \frac{\|R(v(x,t),x)\|}{t} = 0$$

then applies Lemma 1.

3. Starlike Univalent Maps.

Following the usual terminology in the plane, we define a starlike map as follows.

Definition 1. A holomorphic map $f\colon B \to Y$ is starlike if f is univalent, $f(0) = 0$ and $(1 - t)f(B) \subset f(B)$ for all $t \in I$ (thus $f(B)$ is starlike with respect to the origin in the usual sense).

The following theorem characterizes the locally biholomorphic starlike maps by an analytic condition that is equivalent to the usual condition $\mathrm{Re}\,[zf'(z)/f(z)] > 0$ when X is one-dimensional. This result was obtained by Matsuno in 1955 [9] for the case $X = C^n$ with Euclidean norm, it was obtained by Suffridge for the case $X = C^n$ with sup norm in 1970 [15] and for the case of a general Banach space with the boundedness condition on the mapping function $\|Df(x)^{-1}\| \leq M_f(r)$ when $x \in B_r$ in 1973 [16]. Finally, in 1975 [4], Gurganus showed that the boundedness condition was not needed. The final result is as follows.

Theorem 1. *Suppose* $f: B \to Y$ *is locally biholomorphic (i.e.* $Df(x)$ *is a bounded linear map with a bounded inverse for each* $x \in B$*) and* $f(0) = 0$*. Then* f *is starlike if and only if*

(1) $$f(x) = Df(x)(w(x)) \qquad \text{where } w \in \mathcal{M}\ .$$

Outline of Proof. The "only if" part of the theorem follows readily from the observation that the function $F(x,t) = (1-t)f(x)$ satisfies the hypotheses of Lemma 2 with $G(x) = f(x)$. Thus, (1) holds with $w \in \mathcal{N}_0$. However, it then follows that

$$Df(0)(x) = \lim_{\beta \to 0} \frac{1}{\beta} f(\beta x) = \lim_{\beta \to 0} Df(\beta x)\left(\frac{1}{\beta} w(\beta x)\right) = Df(0)(Dw(0)(x))$$

so that $Dw(0)(x) = x$ and we conclude $w \in \mathcal{M}$.

The "if" part of the theorem is not quite as evident. It is true essentially because (1) yields the result that if $v(x,t) = f^{-1}(1-t)f(x))$ $(= x - tw(x) + o(t)$) for fixed $x \in B$ and t near 0, then $\|v(x,t)\|$ is strictly decreasing as a function of t. This allows one to show that if $\|x\| = \|y\|$ and $f(x) = f(y)$ then the inverse images of the curves $\{(1-t)f(x): 0 \leqslant t \leqslant 1\}$ and $\{(1-t)f(y): 0 \leqslant t \leqslant 1\}$ (of course $f(x) = f(y) \Rightarrow$ these are the same segment but in one case, we assume $f^{-1}(f(x)) = x$ and in the other case $f^{-1}(f(y)) = y$) must either be identical or disjoint. Since 0 is on both curves, they are identical, so $x = y$. In order to complete the details of this proof, [16] I required that given r, $0 < r < 1$, there exist $M(r) < +\infty$ such that $\|Df(x)^{-1}\| \leqslant M(r)$ when $x \in B_r$. Gurganus' proof [4] uses an existence and uniqueness theorem for solutions of the initial value problem $\partial v(x,t)/\partial t = -w(v(x,t))$, $v(x,0) = x$. In this case, one concludes $v(x,t) = f^{-1}(e^{-t}f(x))$, $t \geqslant 0$.

Example 3. Let $X = C^2$ with the norm $\|z\|_p = \|(z_1,z_2)\|_p = [|z_1|^p + |z_2|^p]^{1/p}$ where p is fixed $1 < p < \infty$. If $u = (u_1,u_2) \neq (0,0)$, $T(u)$ consists of the one element ℓ_u given by

$$\ell_u(z) = \frac{\dfrac{|u_1|^p}{u_1} z_1 + \dfrac{|u_2|^p}{u_2} z_2}{\|u\|_p^{p-1}}\ .$$

Let

(2) $$f(z) = (z_1 + az_2^2, z_2)$$

so that $f(z) = Df(z)(w(z))$ where $w(z) = (z_1 - az_2^2, z_2)$. By a computation using only elementary calculus, it follows that $w \in \mathcal{M}$ if and only if

$$\frac{p+1}{p-1}\left(\frac{p^2-1}{4}\right)^{1/p} \geqslant |a|\ .$$

In case $p = \infty$ and $\|z\|_\infty = \max(|z_1|, |z_2|)$, if $u = (u_1,u_2) \neq (0,0)$ then $T(u)$ consists of

(i) the functional $\ell(z) = |u_1|z_1/u_1$ if $|u_1| > |u_2|$ or

(ii) the functional $\ell(z) = |u_2|z_2/u_2$ if $|u_2| > |u_1|$ or

(iii) all functionals of the form $\ell_t(z) = t(|u_1|z_1/u_1) + (1-t)(|u_2|z_2/u_2)$, $0 \leqslant t \leqslant 1$ if $|u_2| = |u_1|$. In this case $w = (w_1,w_2) \in \mathcal{N}$ if and only if $\operatorname{Re} w_j(z)/z_j > 0$ whenever $\|z\| = |z_j|$, $j = 1,2$. The function f given by (2) is starlike if and only if $|a| \leq 1$.

If case $p = 1$ and $\|z\| = |z_1| + |z_2|$, if $u = (u_1,u_2) \neq (0,0)$ then $T(u)$ consists of

(i) the functional $\ell(z) = (|u_1|/u_1)z_1 + (|u_2|/u_2)z_2$ if $u_1 \neq 0$ and $u_2 \neq 0$ or

(ii) all functionals of the form $\ell(z) = (|u_1|/u_1)z_1 + \alpha z_2$, $|\alpha| \leq 1$ if $u_2 = 0$ or

(iii) all functionals of the form $\ell(z) = \alpha z_1 + (|u_2|/u_2)z_2$, $|\alpha| \leq 1$ if $u_1 = 0$.

In this case, the function f given by (2) is starlike if and only if $|z_1| - |a|\,|z_2|^2 + |z_2| \geq 0$ when $|z_1| + |z_2| < 1$. Therefore, the condition for starlikeness in this case is $|a| \leq 1$.

Example 4. Let $X = \mathbb{C}^n$ with sup norm. Given $u \neq 0$, $u \in X$, $T(u)$ consists of all functionals of the form

$$\ell(z) = \sum_{|u_j| = \|u\|} t_j \frac{|u_j|}{u_j} z_j$$

where $t_j \geq 0$ for each j and $\sum_{|u_j|=\|u\|} t_j = 1$. Then $w\colon B \to X$ is in the class N provided $w(0) = 0$ and $\operatorname{Re}(w_j(z)/z_j) > 0$ whenever $\|z\| = |z_j| > 0$. Consider $f\colon B \to \mathbb{C}^n$ given by $f(z_1,z_2,\ldots,z_n) = (z_1, (1-z_1)z_2, (1-z_1)z_3, \ldots, (1-z_1)z_n)$. Then $f(z) = Df(z)(w(z))$ where $w_1(z) = z_1$ and $w_j(z) = z_j/(1-z_1)$, $2 \leq j \leq n$. Therefore, f is starlike. This example is interesting in that every boundary point of the form $(1,z_2,\ldots,z_n)$ has the image $(1,0,0,\ldots,0)$. Thus an $(n-1)$-dimensional piece of the boundary maps to a single point yet the function is univalent (and even starlike) in the open unit ball B. A similar result holds for $X = \ell^\infty$ = Banach space of bounded sequences with sup norm but the details are more difficult to verify (see [16, p. 581]).

Example 5. $X = \mathbb{C}^2$ with norm $\|z\|_p$ as given in Example 3 (here p is fixed, $1 \leq p \leq \infty$). Using the information concerning $T(z)$ given in Example 3, it is not difficult to show that the functions

(i)
$$f(z) = \left(\frac{z_1}{(1-z_1)(1-z_2)}, \frac{z_2}{(1-z_1)(1-z_2)}\right)$$

and

(ii)
$$f(z) = \left(\frac{z_1}{1-az_1z_2}, \frac{z_2}{1-az_1z_2}\right), \qquad a = 4^{1/p}$$

are starlike. In (i), (1) holds with w satisfying

$$w_j = z_j\left(\frac{1-z_1z_2}{(1-z_1)(1-z_2)}\right) = \frac{z_j}{2}\left(\frac{1+z_1}{1-z_1} + \frac{1+z_2}{1-z_2}\right), \qquad j = 1,2$$

while in (ii), (1) holds with $w_j = z_j\,((1+az_1z_2)/(1-az_1z_2))$, $j = 1,2$.

Example 6. Let X be a complex inner-product space with inner product $\langle\ ,\ \rangle$ and, as usual, $\|x\|^2 = \langle x,x\rangle$. Suppose $f(z) = z + a_2z^2 + \cdots$ is a complex valued starlike function in the unit disk $\{|z| < 1\}$ of the complex plane. Let $x_0 \in X$ satisfy $\|x_0\| = 1$ and define $F\colon B \to X$ by $F(x) = (1/\langle x,x_0\rangle)f(\langle x,x_0\rangle)x$ ($= x$ if $\langle x,x_0\rangle = 0$). If $u \neq 0$, $u \in X$, then $T(u)$ consists of the functional ℓ_u defined by $\ell_u(x) = \langle x, u/\|u\|\rangle$ and $w \in N$ if and only if $w\colon B \to X$ is holomorphic, $w(0) = 0$ and $\operatorname{Re}\langle w(x),x\rangle > 0$, $0 < \|x\| < 1$. We have

$$DF(x)(\cdot) = -\frac{\langle\cdot,x_0\rangle}{\langle x,x_0\rangle^2} f(\langle x,x_0\rangle)x + \frac{\langle\cdot,x_0\rangle}{\langle x,x_0\rangle} f'(\langle x,x_0\rangle)x + \frac{1}{\langle x,x_0\rangle} f(\langle x,x_0\rangle)I$$

where I is the identity map. Setting

$$w(x) = \frac{f(\langle x,x_o\rangle)}{\langle x,x_o\rangle f'(\langle x,x_o\rangle)}\, x ,$$

it is clear that $w \in M$ (by the analytic condition for starlikeness of f). Further, $F(x) = DF(x)(w(x))$ and hence F is starlike.

4. Convex Univalent Maps. We now turn to a study of conditions for convexity. Many of the results stated here are in [16] and complete proofs will not be given here. We do give a number of examples and some conjectures. It is convenient to make the following definition.

Definition 2. If $f\colon B \to Y$ is a biholomorphic map of B onto a convex domain, we say that f is convex.

We begin with the following lemma.

Lemma 3. *If* $f\colon B \to Y$ *is convex then* $f(B_r)$ *is convex for each* r, $0 < r \leqslant 1$ [16, Lemma 4].

Proof. Fix $u,v \in B$, $\|v\| \leqslant \|u\|$ and fix t, $0 < t < 1$. Let $\ell \in T(u)$ and set $f(h(x)) = tf(x) + (1-t)f(\frac{\ell(x)}{\|u\|}v)$. Applying Schwarz's lemma to h, we conclude $\|h(u)\| \leqslant \|u\|$ and the lemma is proved.

In the one-dimensional case, a function $f\colon \{|z| < 1\} \to \mathbf{C}$ is convex if and only if $f'(0) \neq 0$ and $\operatorname{Re}\,[zf''(z)/f'(z) + 1] > 0$ when $|z| < 1$. The analogous condition in higher dimensions (dimension 2 or greater) is necessary but not sufficient for convexity.

Theorem 2. *If* $f\colon B \to Y$ *is convex then*

$$D^2f(x)(x,x) + Df(x)(x) = Df(x)(w(x)) \tag{3}$$

where $w \in M$ [16, Theorem 3].

The proof consists of applying Lemma 2 to the function $F(x,t) = \tfrac{1}{2}[f(e^{i\sqrt{t}}\,x) + f(e^{-i\sqrt{t}}\,x)]$.

Recall that in the plane, (3) together with the condition $f'(0) = 1$ implies f is convex and $zf'(z)$ is starlike. Example 7 below shows if X has dimension 2, the analogous result is not necessarily true. In fact, (3) together with $Df(0) = I$ (here we assume $Y = X$ and I is the identity) does not even imply $Df(x)(x)$ is univalent.

Example 7. Let $X = \mathbf{C}^2 = Y$ with norm, $\|z\| = \|(z_1,z_2)\| = (|z_1|^p + |z_2|^p)^{1/p}$ where p is fixed, $1 \leqslant p \leqslant 2$. Define $f\colon B \to X$ by $f(z_1,z_2) = (z_1 + az_1z_2, z_2)$. Then $Df(z)(z) = (z_1 + 2az_1z_2, z_2)$ and (3) becomes $(z_1 + 4az_1z_2, z_2) = ((1 + az_2)w_1 + az_1w_2, w_2)$ so that $w_2 = z_2$ and $w_1 = z_1((1 + 3az_2)/(1 + az_2))$. Clearly, $Df(z)(z)$ is not univalent in B if $|a| > \tfrac{1}{2}$ (for then one can choose $z_2 = -1/2a$ and let z_1 be arbitrary, $|z_1|^p < 1 - (1/2|a|)^p$ to obtain the image $(0,z_2)$). The condition which w must satisfy when $|a| \leqslant 1$ in order that $w \in M$ is $|z_1|^p \operatorname{Re}((1 + 3az_2)/(1 + az_2)) + |z_2|^p \geqslant 0$ or (since $|a| \leqslant 1$),

$$|z_1|^p \frac{1 - 3|a|\,|z_2|}{1 - |a|\,|z_2|} + |z_2|^p \geqslant 0 , \qquad (z_1,z_2) \in B .$$

By elementary calculus, it follows that $w \in M$ if and only if $\frac{p+1}{3p}\left(\frac{2p+1}{3}\right)^{1/p} \geqslant |a|$. However, $\frac{p+1}{3p}\left(\frac{2p+1}{3}\right)^{1/p} > \frac{1}{2}$ when $1 \leqslant p \leqslant 2$ and it follows that for some a, (3) holds with $w \in M$ and $Df(z)(z)$ is not univalent in B.

The condition (3) can readily be shown to imply the geometric property that for small $\epsilon > 0$, the set $\{f(\alpha x): |\alpha| \leqslant 1, 0 < |\alpha - 1| < \epsilon\}$ lies in the half-space $\{y \in Y: \text{Re}\, \ell \circ (Df(x))^{-1}(y) < \text{Re}\, \ell \circ (Df(x)^{-1}(f(x))\}$ determined by the hyperplane "tangent" to $f(B_{\|x\|})$ at $f(x)$. To see this, set $1 - \alpha = \rho e^{i\varphi}$, ρ small and positive, $\cos\varphi \geqslant \rho/2$ (so that $|\alpha| \leqslant 1$). Then

$$f(\alpha x) = f(x + (\alpha - 1)x) = f(x) - \rho e^{i\varphi} Df(x)(x) + \tfrac{1}{2}\rho^2 e^{2i\varphi} D^2 f(x)(x,x) + o(\rho^2).$$

Consider first $\cos\varphi \geqslant \sqrt{\rho}$. In this case, $\text{Re}\, \ell \circ Df(x)^{-1}(f(\alpha x)) = \text{Re}\, \ell \circ Df(x)^{-1}(f(x)) - \rho\cos\varphi\|x\| + O(\rho^2)$ $< \ell \circ Df(x)^{-1}(f(x))$ when ρ is sufficiently small. In case $\rho/2 \leqslant \cos\varphi < \sqrt{\rho}$,

$$\begin{aligned}
&\text{Re}\, \ell \circ Df(x)^{-1}(f(\alpha x)) \\
&= \text{Re}\, \ell \circ (Df(x))^{-1}(f(x)) - \text{Re}\, \ell \circ Df(x)^{-1}[(\rho\cos\varphi + i\sin\varphi)Df(x)(x) + (\rho^2\cos^2\varphi - \tfrac{1}{2}\rho^2 \\
&\qquad + i\rho^2 \sin\varphi\cos\varphi)D^2 f(x)(x,x)] + o(\rho^2) \\
&= \text{Re}\, \ell \circ Df(x)^{-1}(f(x)) - \frac{\rho^2}{2}\text{Re}\, \ell \circ Df(x)^{-1}[Df(x)(x) + D^2 f(x)(x,x)] - \rho(\cos\varphi - \frac{\rho}{2})\|x\| + o(\rho^2) \\
&< \text{Re}\, \ell \circ Df(x)^{-1}(f(x))
\end{aligned}$$

when ρ is sufficiently small (assuming that (3) holds with $w \in M$). It seems unlikely to me that this geometric property can hold for all $x \neq 0$ if $\text{Re}\, \ell \circ (Df(x))^{-1}(f(x)) = 0$ (i.e. the "tangent" hyperplane referred to above passes through the origin) for some x. This leads to the following conjecture.

Conjecture 1. If $f\colon B \to Y$ is holomorphic, $f(0) = 0$ and (3) holds for some $w \in M$ then f is starlike.

In the complex plane, an analytic function $g\colon \{|z| < 1\} \to \mathbf{C}$ is convex if and only if $f(z) = zg'(z)$ is starlike. This relation does not hold in higher dimensions as we will see below. However, we do conjecture the following.

Conjecture 2. If $f\colon B \to Y$ is starlike and $g(x) = \int_0^1 \frac{1}{t} f(tx)dt$ (so $Dg(x)(x) = f(x)$) then g is starlike.

The following example shows that if $X = \mathbf{C}^2$ with $\|z\| = \|(z_1,z_2)\| = (|z_1|^p + |z_2|^p)^{1/p}$ where $1 \leqslant p < \infty$ and $f(z) = (f_1(z_1), f_2(z_2))$ where f_1 and f_2 are convex in the plane, the map f may still fail to be convex. This result is rather surprising and as we will see, the requirement that f(B) be convex is much more restrictive than one might at first expect.

Example 8. Setting $X = \mathbf{C}^2$ with the p-norm, $1 \leqslant p < \infty$, define $f\colon B \to X$ by $f(z_1,z_2) = (z_1/(1-z_1), z_2)$. Then $tf(z) + (1-t)f(w) = f(u) \Rightarrow u_1 = (tz_1 + (1-t)w_1 - z_1w_1)/(1 - (1-t)z_1 - tw_1)$ and $u_2 = tz_2 + (1-t)w_2$. Setting $w = (r,0)$ and $z = (r-\epsilon, (r^p - (r-\epsilon)^p)^{1/p})$ where $1 > r > 0$, $r > \epsilon > 0$ we find that

$$\|u\|^p = \left[\frac{t(r-\epsilon)+(1-t-(r-\epsilon))r}{1-(1-t)(r-\epsilon)-tr}\right]^p + t^p(r^p-(r-\epsilon)^p)\,.$$

Hence

$$\left.\frac{d\|u\|^p}{dt}\right|_{t=1} = p(r-\epsilon)^{p-1}\left(\frac{(-\epsilon)(1-r+\epsilon)}{1-r}\right) + p(r^p-(r-\epsilon)^p)\,.$$

Now let $\epsilon = k(1-r)$ where k is large and let r be near 1 so that ϵ is small. It is clear that one can make $d\|u\|^p/dt\,|_{t=1} < 0$ so $\|u\| > r$ for some t, $0 < t < 1$. Then f cannot be convex.

The following theorem [16, Theorems 4 and 5] gives the necessary and sufficient condition for convexity

Theorem 3. *Suppose* $f\colon B \to Y$ *is locally biholomorphic and set*

(4) $$f(x) - f(y) = Df(x)(w(x,y))\,, \qquad x,y \in B\,.$$

Then f *is convex if and only if* $\operatorname{Re} x^*(w(x,y)) > 0$ *whenever* $\|y\| < \|x\|$ *and* $x^* \in T(x)$.

The condition given in the theorem is a condition which forces $f(B)$ to be starlike with respect to each of its points. The proof uses this idea and is similar to the proof of Theorem 1.

Example 9. Let f be defined in C^2 by $f(z_1,z_2) = (z_1 + az_2^2, z_2)$. By applying Theorem 3, it is relatively easy to show that:

(i) if $\|z\| = |z_1| + |z_2|$, f is convex if and only if $a = 0$;

(ii) if $\|z\|^2 = |z_1|^2 + |z_2|^2$, f is convex if and only if $|a| \leqslant ½$ and

(iii) if $\|z\| = \max(|z_1|,|z_2|)$, f is convex if and only if $a = 0$.

Theorem 3 is usually rather tedious to apply and hence the following result is sometimes helpful [16, Theorem 6].

Theorem 4. *Let* $f\colon B \to Y$ *be convex,* $x \in B$, $x \neq 0$, *and let* $\ell \in T(x)$. *Then the hyperplane* $\{y \in Y: \operatorname{Re} \ell \circ Df(x)^{-1}(y) = \operatorname{Re} \ell \circ Df(x)^{-1}(f(x))\}$ *is a supporting hyperplane for the convex set* $f(B_{\|x\|})$. *If* $y \neq 0$, $\operatorname{Re} \ell(y) = 0$ *and* $\|x + ty\| = \|x\|$ *for* $0 < t < t_0$ *then* $\operatorname{Re} \ell \circ Df(x)^{-1}(f(x+ty)) = \operatorname{Re} \ell \circ Df(x)^{-1}(f(x))$ *for* $0 < t < t_0$ *(i.e.* $f(x+ty)$ *lies in the supporting hyperplane described above).*

Proof. Assume $\|y\| < \|x\|$, $x \in B$ and apply Theorem 3. We have $0 < \operatorname{Re} \ell \circ Df(x)^{-1}(f(x) - f(y))$ and this implies that all points in $f(B_{\|x\|})$ lie on the same side of the hyperplane described in the theorem. Thus, the first part of the theorem is proved.

Now assume the hypotheses of the second part of the theorem. Expanding $f(x+ty)$ in a power series about x, we have

$$f(x+ty) = f(x) + tDf(x)(y) + \cdots + \frac{1}{N!}t^N D^N f(x)(y^N) + o(t^N)\,.$$

Thus,

$$\begin{aligned} w(x,x+ty) &= Df(x)^{-1}(f(x) - f(x+ty)) \\ &= (Df(x))^{-1}(-tDf(x)(y) - ½t^2D^2f(x)(y,y) - \cdots - \frac{1}{N!}t^N D^N f(x)(y^N) + o(t^N))\,. \end{aligned}$$

Using the fact that $\operatorname{Re}\ell(y) = 0$ and writing $A_k = Df(x)^{-1} \circ D^k f(x)$ we have

$$0 \leqslant \operatorname{Re}\ell(w(x,x+ty)) = -\sum_{k=2}^{\infty} \frac{1}{k!} t^k \operatorname{Re}\ell(A_k(y^k)) .$$

If N is a minimum such that $\operatorname{Re}\ell(A_N(y^N)) \neq 0$, then $0 \leqslant -\frac{1}{N!} t^N \operatorname{Re}\ell(A_N(y^N)) + o(t^N)$ and we conclude $\operatorname{Re}\ell(A_N(y^N)) < 0$. Replacing x by λx, y by λy and ℓ by ℓ_λ where $\ell_\lambda(u) = \ell((|\lambda|/\lambda)u)$, $0 < |\lambda| < 1/\|x\|$ we have $\ell_\lambda \in T(\lambda x)$ and hence $g(\lambda) = \operatorname{Re}\ell(\lambda^{N-1} A_N(\lambda x)(y^N)) < 0$ for all such λ. However, g is harmonic in $0 < |\lambda| < 1/\|x\|$ and assumes the value 0 at $\lambda = 0$. By the maximum principle, $g(\lambda) \equiv 0$ contradicting the choice of N. This completes the proof.

Notice that the theorem says that if any segment $\{x + ty: 0 \leqslant t \leqslant t_0\}$ lies on the boundary of the ball $B_{\|x\|}$ (which occurs for example in ℓ^1 and ℓ^∞) then the curve $\{f(x+ty): 0 \leqslant t \leqslant t_0\}$ lies in some supporting hyperplane of $f(B_{\|x\|})$ at the point $f(x)$.

From Theorem 4, one can prove the following two theorems which are somewhat surprising results. They show that the requirement that $f(B)$ be convex is indeed very restrictive (at least in some spaces). The proofs in [16, pp. 585–588] are rather tedious and will not be given here. There probably are more intuitive proofs of these results but they have eluded me.

Theorem 5. *If* $X = \ell^1$ *(the space of summable complex sequences) and* $f: B \to Y$ *is convex, then* $f(x) - f(0)$ *is linear.*

Theorem 6. *If* $X = \ell^\infty$ *(the space of bounded complex sequences) and* $f: B \to Y$ *is convex, then* $f(x) - f(0) = Df(0)(g_1(x_1), g_2(x_2), \ldots)$ *where* $g_k(x_k) = x_k + a_{2k}x_k^2 + \cdots$ *maps the disk* $|x_k| < 1$ *onto a convex domain.*

The finite dimensional versions of Theorems 5 and 6 with the same norms are also true and Example 9 illustrates this.

In some sense, Theorem 4 shows that if f is holomorphic and convex on B then the boundary of $f(B_{\|x\|})$ has the same "flatness" properties as the boundary of $B_{\|x\|}$. Notice, for example, that if $X = C^2$ with $\|(z_1,z_2)\| = |z_1| + |z_2|$ and $f(z) = z + g(z)$ is holomorphic and univalent on B where g consists of terms of degree 2 or more, then $1 > \|z\| > 0 \Rightarrow f(B_{\|z\|})$ is not convex unless $g(z) \equiv 0$.

5. Close-to-Convexity and Close-to-Starlikeness in Banach Spaces. The term close-to-convex was first used by Kaplan in 1952 [7]. Some ideas related to this concept had been studied previous to this time (e.g. [1] and [2]) but Kaplan studied the geometry of such maps in some detail. Kaplan's definition says that a function $f(z) = z + a_2z^2 + \cdots$, analytic in $\{|z| < 1\}$ is close-to-convex if there exists φ convex such that

(5) $$\operatorname{Re}\left(\frac{f'(z)}{\varphi'(z)}\right) > 0 \qquad \text{when } |z| < 1 .$$

Equivalently, f is close-to-convex if there exists g starlike, $g(z) = a_1z + a_2z^2 + \cdots$ (a_1 is not necessarily real but $\operatorname{Re} a_1 > 0$) such that

(6) $$\operatorname{Re}\left(\frac{zf'(z)}{g(z)}\right) > 0 \qquad \text{when } |z| < 1 .$$

Since the relation: φ is convex if and only if $z\varphi'(z)$ is starlike, does not hold in higher dimensions, the extensions of these two conditions to higher dimensions will probably not be equivalent. The first of these conditions can be extended in a relatively straightforward way to higher dimensions.

Theorem 7. *If* U *is an open convex subset of* X *and* $f: U \to X$ *is holomorphic and satisfies* $Df(u)(\cdot) \in N$ *for each* $u \in U$ *(i.e. given* $u \in U$, $x \in B$, $x \neq 0$ *and* $\ell \in T(x)$ *we have* $\operatorname{Re} \ell(Df(u)(x)) > 0$*) then* f *is univalent in* U.

Proof. Let $x_1, x_2 \in U$, $x_2 \neq x_1$ and $\ell \in T(x_2 - x_1)$. Since U is convex, $\{(1-t)x_1 + tx_2 : 0 \leq t \leq 1\} \subset U$. Differentiating $f((1-t)x_1 + tx_2)$ with respect to t, we find

$$f(x_2) - f(x_1) = \int_0^1 Df((1-t)x_1 + tx_2)(x_2 - x_1)dt .$$

Therefore, $\operatorname{Re} \ell(f(x_2) - f(x_1)) = \int_0^1 \operatorname{Re}(\ell(Df((1-t)x_1 + tx_2)(x_2 - x_1)))dt > 0$ so that $f(x_2) \neq f(x_1)$.

Theorem 8. *If* $f: B \to X$ *is analytic and satisfies* $Df(u) \circ D\varphi(u)^{-1}(\cdot) \in N$ *where* $\varphi: B \to X$ *is convex then* f *is univalent in* B.

Proof. Apply Theorem 7 to $f \circ \varphi^{-1}$.

Definition 3. A holomorphic function $f: B \to Y$ is close-to-convex if there exists $\varphi: B \to X$, φ convex, such that $Df(u) \circ D\varphi(u)^{-1}(\cdot) \in N$ for all $u \in B$.

Theorem 8 says that close-to-convex functions are univalent.

Example 10. Consider $f: \mathbf{C}^2 \to \mathbf{C}^2$ given by $f(z_1,z_2) = (z_1 + az_2^2, z_2)$.
(i) Using the sup norm and restricting f to $B = \{(z_1,z_2): |z_1| < 1, |z_2| < 1\}$ we see that f is close-to-convex by choosing

$$\varphi(z) = \begin{cases} z & \text{if } a = 0 \\ (z_1, 2|a|z_2) & \text{if } a \neq 0 \end{cases}$$

Thus, f is close-to-convex for all a, starlike if $|a| \leq 1$ (Example 3) and convex if $a = 0$ (Example 9).
(ii) Using the Euclidean norm, f is close-to-convex for all a (choose $\varphi(z) = (\epsilon z_1, z_2)$ where ϵ is 1 if $|a| \leq 1/2$, $\epsilon = 1/4|a|^2$ if $|a| \geq 1/2$), it is starlike when $|a| \leq 3\sqrt{3}/2$ (Example 3) and convex when $|a| \leq 1/2$ (Example 9).

In [11], Pfaltzgraff and I extended (6) to finite dimensional spaces.

Definition 4. Given a holomorphic function $f: B \to X$ such that $Df(0) = I$, f is close-to-starlike if there exists $h \in M$ and g starlike such that $Df(x)(h(x)) = g(x)$.

This definition requires that $Dg(0) = I$. This requirement is not really essential but was merely a convenience. The term "close-to-starlike" was used since a starlike function g is not related to a convex function in a simple way in higher dimensions. Our results are quite geometric and involve the concept of a subordination chain (see [10] and [12]). A subordination chain is a function $F: B \times [0,\infty) \to X$ such that for each $t \in [0,\infty)$, $F(\cdot,t)$ is holomorphic, $DF(0,t) = e^t I$ and such that there exist Schwarz functions $v(x,s,t)$, $0 \leq s \leq t$, $x \in B$ (i.e. for $0 \leq s \leq t$ and $x \in B$ we have $\|v(x,s,t)\| \leq \|x\|$) that satisfy $F(x,s) = F(v(x,s,t),t)$. A subordination chain $F(x,t)$ is a univalent subordination chain if for each $t \in [0,\infty)$, $F(\cdot,t)$ is univalent in B.

The following theorem shows that in the finite dimensional case, close-to-starlike functions are univalent (see [11]).

Theorem 9. *Suppose* X *is finite dimensional,* $f\colon B \to X$ *is locally biholomorphic,* $Df(0) = I$, *that* $g\colon B \to X$ *is starlike and* $Dg(0) = I$. *Then* $F(x,t) = f(x) + (e^t - 1)g(x)$ *is a univalent subordination chain if and only if* f *is close-to-starlike relative to the starlike function* g.

The proof of this result is rather complicated and will not be given here. The result is probably true in the infinite dimensional case but our proof uses finite dimensionality in an essential way. The geometry is made more clear by the following result [11] which is similar to Lewandowski's result for close-to-convex functions in the plane.

Theorem 10. *If* f *is close-to-starlike relative to the starlike function* g, *then for each* r, $0 < r < 1$, *the complement in* X *of* $f(B_r)$ *is the union of nonintersecting rays.*

Proof. It is clear from Theorem 9 that for fixed r, $0 < r < 1$, the rays $L(x) = \{f(x) + tg(x)\colon t \geq 0\}$, as x varies over the sphere S_r, are disjoint and fill up the complement of $f(B_r)$.

Example 11. Suppose $X = \ell^p$, $1 \leq p < \infty$ and $f\colon B \to X$ and $g\colon B \to X$ are given by $f(x) = (f_1(x_1), f_2(x_2), \ldots)$ and $g(x) = (g_1(x_1), g_2(x_2), \ldots)$ where f_j is close-to-convex (in the usual sense) relative to the starlike function g_j, $j = 1,2,\ldots$. If $x \in B$, $x \neq 0$ and $x^* \in T(x)$, then

$$x^*((Df(x))^{-1}(g(x))) = \frac{1}{\|x\|^{p-1}} \sum_{j=1}^{\infty} |x_j|^p \frac{g_j(x_j)}{x_j f_j'(x_j)}$$

and hence $Df(x)(w(x)) = g(x)$ where $w \in N$. Thus, f is close-to-starlike (except possibly for the normalization $D_w(0) = I$).

Example 12. Let X be an inner-product space with inner product $\langle\cdot,\cdot\rangle$. Suppose $f(z) = z + a_2 z^2 + \cdots$ is analytic in the unit disk of the complex plane and $\operatorname{Re}(zf'(z)/g(z)) > 0$ where g is starlike. Let $x_0 \in X$ satisfy $\|x_0\| = 1$ and define $F(x) = (1/\langle x,x_0\rangle) f(\langle x,x_0\rangle)x$ $(= x$ if $\langle x,x_0\rangle = 0)$ and $G(x) = (1/\langle x,x_0\rangle) g(\langle x,x_0\rangle)x$. From Example 6, G is starlike. Further, $DF(x)(H(x)) = G(x)$ where $H(x) = (g(\langle x,x_0\rangle)/\langle x,x_0\rangle f'(\langle x,x_0\rangle))x$ and it is clear that $H \in N$. Thus, F is close-to-starlike (again modulo the requirement $DH(0) = I$).

6. Spirallike Functions. A function $f(z) = z + a_2 z^2 + \cdots$ that is analytic in the unit disk $\{|z| < 1\}$ is said to be spirallike if $\operatorname{Re} e^{i\alpha} zf'(z)/f(z) > 0$ for some fixed α, $|\alpha| < \pi/2$ and all z, $|z| < 1$. Spirallike functions are univalent [14]. The terminology is derived from the fact that if f is spirallike relative to $e^{i\alpha}$ (i.e. $\operatorname{Re} e^{i\alpha} zf'(z)/f(z) > 0$) then for each $t \geq 0$, the function $F(z,t) = e^{-te^{i\alpha}} f(z)$ is subordinate to f. Actually, for fixed z, $|z| = r < 1$, the spiral $\{e^{-te^{i\alpha}} f(z)\colon t \text{ is real}\}$ intersects the curve $\{f(re^{i\theta}), 0 \leq \theta < 2\pi\}$ in exactly one point. In higher dimensions, we define spirallikeness as follows.

Definition 5. Suppose $f: B \to X$ is a biholomorphic map of B onto an open set $U \subset X$, $f(0) = 0$, $Df(0) = I$ and suppose there exists a bounded linear map $A: X \to X$ and $\epsilon > 0$ such that

$$(7) \qquad x \in X, \quad \|x\| = 1, \quad x^* \in T(x) \Rightarrow \operatorname{Re} x^*(A(x)) \geqslant \epsilon .$$

Then f is spirallike relative to A if $e^{-tA}f(B) \subset f(B)$ for all $t \geqslant 0$ where $e^{-tA} = \Sigma_{k=0}^{\infty} \frac{1}{k!} t^k A^k$ (compare [4, Definition 3 and Theorem 4]).

Theorem 11. *Suppose* $A: X \to X$ *is a bounded linear map that satisfies* (7) *and* $f: B \to X$ *is locally biholomorphic,* $f(0) = 0$, $Df(0) = I$. *Then* f *is spirallike if and only if*

$$(8) \qquad Af(x) = Df(x)(w(x))$$

for some $w \in N$.

Proof. Apply Lemma 2 to $F(x,t) = e^{-tA}f(x)$ to obtain the "only if" part of the theorem.

To obtain the converse, let $v(x,t)$ be the unique solution of the initial value problem

$$(9) \qquad \frac{\partial v_x}{\partial t} = -w(v_x), \qquad v_x(0) = x$$

where w satisfies (8), $w \in N$ [10, Theorem 2.1] (also see [4, Lemma 5]). From (8) we have $A \circ Df(0) = Df(0)(Dw(0))$ so that $Dw(0) = A$. By the remarks in [4, p. 403, formula 27], $\|v(x,t)\| \to 0$ as $t \to +\infty$. Finally, using [4, Corollary 2] we conclude f is univalent. Setting $u(x,t) = f^{-1}(e^{-tA}f(x))$, $x \in B$, t near 0, we find that u is a solution of the initial value problem (8). Thus, $u(x,t) = v(x,t)$ is a Schwarz function and $e^{-tA}f(B) \subset f(B)$ for each $t \geqslant 0$. This completes the proof.

Example 13. Let $X = \ell^p$ and suppose $f: B \to X$ satisfies $f(x) = (f_1(x_1), f_2(x_2), \ldots)$ where $f_j(x_j)$ is spirallike relative to $e^{i\alpha_j}$, $|\alpha_j| < \pi/2$, $j = 1,2,\ldots$. Then f is spirallike relative to A where $A(x) = (e^{-i\alpha_1}x_1, e^{-i\alpha_2}x_2, \ldots)$.

Example 14. Let $X = \mathbf{C}^2$ with p norm $1 \leqslant p \leqslant \infty$ and let

$$f(z) = \left(\frac{z_1}{(1 + e^{i\alpha}z_1z_2)^{(1+e^{-i\alpha})/2}} , \frac{z_2}{(1 + e^{i\alpha}z_1z_2)^{(1+e^{-i\alpha})/2}} \right)$$

where $|\alpha| < \pi$ and principal branches are used for the powers. If A is the linear map

$$e^{-i\alpha/2} I = \begin{pmatrix} e^{-i\alpha/2} & 0 \\ 0 & e^{-i\alpha/2} \end{pmatrix}$$

and

$$w(z) = \left(\frac{z_1 e^{-i\alpha/2}(1 + e^{i\alpha}z_1z_2)}{1 - z_1z_2} , \frac{z_2 e^{-i\alpha/2}(1 + e^{i\alpha}z_1z_2)}{1 - z_1z_2} \right)$$

then $w \in N$ and $Af(z) = Df(z)(w(z))$. Hence f is spirallike.

References

1. J. W. Alexander, Functions which map the interior of the unit circle upon simple regions, *Annals of Math.* **17** (1915), 12–22.

2. M. Biernacki, Sur la représentation conforme des domaines linéairement accessible, *Prace Frat. Fiz.* **44** (1937), 293–314.

3. S. Bochner and W. T. Martin, **Several Complex Variables**, Princeton University Press, 1948.

4. K. R. Gurganus, Φ-like holomorphic functions in C^n and Banach spaces, *Trans. Amer. Math. Soc.* **205** (1975), 389–406.

5. L. A. Harris, Schwarz's Lemma in normed linear spaces, *Proc. Natl. Acad. Sci. U.S.A.* **64**(4), (1969), 1014–1017.

6. E. Hille and R. S. Phillips, Functional Analysis and Semigroups, *Amer. Math. Soc. Colloq. Publ.* **31** (1957).

7. W. Kaplan, Close-to-convex schlicht functions, *Michigan Math. J.* **1** (1952), 169–185.

8. Z. Lewandowski, Sur l'identité de certaines classes de fonctions univalentes, I and II, *Ann. Univ. Mariae Curie-Sklodowska,* Sect. A, **12** (1958), 131–146 and **14** (1960), 19–46.

9. T. Matsuno, Starlike theorems and convex-like theorems in the complex vector space, *Sci. Rep. Toko, Kyoiku Daigaku,* Sect. A, **5** (1955), 88–95.

10. J. A. Pfaltzgraff, Subordination chains and univalence of holomorphic mappings on C^n, *Math. Ann.* **210** (1974), 55–68.

11. J. A. Pfaltzgraff and T. J. Suffridge, Close-to-starlike holomorphic functions of several variables, *Pac. J. Math.* **57** (1975), 271–279.

12. Ch. Pommerenke, Über die Subordination analytischer Funktionen, *J. Reine Angew. Math.* **218** (1965), 159–173.

13. M. S. Robertson, Applications of the subordination principle to univalent functions, *Pac. J. Math.* **11** (1961), 315–324.

14. L. Spaček, Contibution à la theorie des fonctions univalentes, *Časopsis Pěct. Mat.* **62** (1932), 12–19.

15. T. J. Suffridge, The principle of subordination applied to functions of several variables, *Pac. J. Math.* **33** (1970), 241–248.

16. ————, Starlike and convex maps in Banach spaces, *Pac. J. Math.* **46** (1973), 575–589.

University of Kentucky
Lexington, Kentucky 40506